CAMBRIDGE LIBRAR

Books of enduring sch

Archaeology

The discovery of material remains from the recent or the ancient past has always been a source of fascination, but the development of archaeology as an academic discipline which interpreted such finds is relatively recent. It was the work of Winckelmann at Pompeii in the 1760s which first revealed the potential of systematic excavation to scholars and the wider public. Pioneering figures of the nineteenth century such as Schliemann, Layard and Petrie transformed archaeology from a search for ancient artifacts, by means as crude as using gunpowder to break into a tomb, to a science which drew from a wide range of disciplines - ancient languages and literature, geology, chemistry, social history - to increase our understanding of human life and society in the remote past.

Early Adventures in Persia, Susiana, and Babylonia

Sir Austen Henry Layard (1817–94) was one of the leading British archaeologists of the Victorian period. His excavations at Nimrud and Nineveh led to important discoveries about ancient Mesopotamia, particularly about the Assyrian civilisation, and his popular books such as *Nineveh and its Remains* (1849) brought archaeology to a wide audience. This two-volume work, first published in 1887, tells the story of an 'adventurous journey' Layard had made over forty years earlier, in 1840–2. He learnt Arabic and Persian and travelled widely, even among tribal peoples notorious for their lawlessness. These included the mountain-dwelling Bakhtiyari, who were under threat from the Persian ruler. Volume 1 describes the ancient sites Layard visited at the start of his journey, his encounters with the authorities in several cities, the warm hospitality he experienced in the household of the Bakhtiyari chiefs, and their customs, including a lion hunt and recitations of poetry.

Cambridge University Press has long been a pioneer in the reissuing of out-of-print titles from its own backlist, producing digital reprints of books that are still sought after by scholars and students but could not be reprinted economically using traditional technology. The Cambridge Library Collection extends this activity to a wider range of books which are still of importance to researchers and professionals, either for the source material they contain, or as landmarks in the history of their academic discipline.

Drawing from the world-renowned collections in the Cambridge University Library, and guided by the advice of experts in each subject area, Cambridge University Press is using state-of-the-art scanning machines in its own Printing House to capture the content of each book selected for inclusion. The files are processed to give a consistently clear, crisp image, and the books finished to the high quality standard for which the Press is recognised around the world. The latest print-on-demand technology ensures that the books will remain available indefinitely, and that orders for single or multiple copies can quickly be supplied.

The Cambridge Library Collection brings back to life books of enduring scholarly value (including out-of-copyright works originally issued by other publishers) across a wide range of disciplines in the humanities and social sciences and in science and technology.

Early Adventures in Persia, Susiana, and Babylonia

Including a Residence Among the Bakhtiyari and Other Wild Tribes Before the Discovery of Nineveh

Volume 1

Austen Henry Layard

CAMBRIDGE UNIVERSITY PRESS

Cambridge, New York, Melbourne, Madrid, Cape Town,
Singapore, São Paolo, Delhi, Tokyo, Mexico City

Published in the United States of America by Cambridge University Press, New York

www.cambridge.org
Information on this title: www.cambridge.org/9781108043427

© in this compilation Cambridge University Press 2012

This edition first published 1887
This digitally printed version 2012

ISBN 978-1-108-04342-7 Paperback

This book reproduces the text of the original edition. The content and language reflect
the beliefs, practices and terminology of their time, and have not been updated.

Cambridge University Press wishes to make clear that the book, unless originally published
by Cambridge, is not being republished by, in association or collaboration with, or
with the endorsement or approval of, the original publisher or its successors in title.

The original edition of this book contains a number of colour plates,
which have been reproduced in black and white. Colour versions of these
images can be found online at www.cambridge.org/9781108043427

EARLY ADVENTURES
IN
PERSIA, SUSIANA, AND BABYLONIA

VOL. I.

Sir A. Henry Layard in Bakhtiyari Costume.

EARLY ADVENTURES
IN
PERSIA, SUSIANA, AND BABYLONIA

INCLUDING A RESIDENCE

AMONG THE BAKHTIYARI AND OTHER WILD TRIBES

BEFORE THE DISCOVERY OF NINEVEH

By SIR HENRY LAYARD, G.C.B.

AUTHOR OF 'NINEVEH AND ITS REMAINS' ETC.: GOLD MEDALLIST OF
THE ROYAL GEOGRAPHICAL SOCIETY

IN TWO VOLUMES—VOL. I.

With Maps and Illustrations

LONDON
JOHN MURRAY, ALBEMARLE STREET
1887

All rights reserved

PRINTED BY
SPOTTISWOODE AND CO., NEW-STREET SQUARE
LONDON

CONTENTS

OF

THE FIRST VOLUME.

———◆———

CHAPTER I.

	PAGE
INTRODUCTORY	1

CHAPTER II.

Leave Jerusalem—Bethlehem—Latins and Greeks—Hebron—Colonel Yusuf Effendi—The bastinado—Sheikh Abu-Dhaouk—Departure for Petra—The tents of Abu-Dhaouk—Savage country—Danger from Arab robbers—Meet Bedouin horsemen—A sacred spot—Enter the mountains—The Wady Musa—Ascend a peak—Rock-cut monuments—Reach Petra—Difficulties with the Arabs—The ruins ... 21

CHAPTER III.

Leave Petra—Danger from Arabs—The Wady Ghor—The Seydi'in Arabs—The Dead Sea—The Mountains of Moab—Plundered by Arabs—Arrival at Kerak—The son of the Mujelli—Recovery of my property—The Christians of Kerak—Its ruins—My Christian host—Sheikh Suleiman-

[6] EARLY ADVENTURES

PAGE

Ibn-Fais—Departure for the Desert—Encampments of Christian Arabs—The Country of Moab—The Beni-Hamideh—Remains of ancient towns—Meshita—An Arab banquet—The tents of Suleiman-Ibn-Fais—Reach Ammon . 64

CHAPTER IV.

The ruins of Ammon—Sheikh Suleiman Shibli—Fortunate escape—A funeral—The plague—Reach Jerash—Description of the ruins—Isaac of Hebron—The plague at Remtheh —Irbid—A Bashi-Bozuk—Cross the Jordan—Maäd—Deserted by my guide—Tiberias—Hyam, a Jew—His generosity—Safed—The Jew Shimoth—Effects of the earthquake—An Arab muleteer and his wife—Start for Damascus—Robbed by deserters—Kaferhowar—Evading the quarantine—Escape from arrest—Arrival at Damascus . . 123

CHAPTER V.

Mr. Consul Wherry—A Turkish bath—Damascus described—An Arab barber-surgeon—Padre Tommaso—Persecution of the Jews—The French Consul—Purchase a mare—Leave Damascus—Cross Anti-Lebanon—The Mutuali—Arrive at Baalbek—An Italian military instructor—The Emir—Meet with an accident—Leave for Beyrout—Cross Lebanon—Reach Beyrout—Journey to Aleppo—Rejoin Mr. Mitford—Leave Aleppo for Baghdad 174

CHAPTER VI.

Leave Baghdad—Join a caravan—Incidents on our march—Village of Yakubiyeh—Kizilrobat—Khanikin—Our travelling-companions—Ruins of Holwan—Meet a French Ambassador—Sculptures at Ser-Puli Zohab—The Ali-Ilahis—Cross the Persian frontier—Kirrind—The Lurs—Reach Kermanshah—Sculptures of Taki Bostan—Persian fanaticism—Difficulties at Kermanshah—The Governor—A Mùnshi—Continue our journey—The sculptures of Bisutun

CONTENTS OF THE FIRST VOLUME [7]

PAGE

—The Shah's camp—The Hakim-Bashi—The Minister for Foreign Affairs—The camp raised—The Shah—Reach Hamadan—French officers—The Prime Minister—Hussein Khan—Difficulty in obtaining firman—The Baron de Bode—Cuneiform inscriptions—Separate from Mr. Mitford   201

CHAPTER VII.

Leave Hamadan—My mehmandar—Douletabad—A Persian palace—Kala Khalifa—Burujird—Khosrauabad—Difficulties of the journey—A village chief—The Bakhtiyari—One of their chiefs—Renounce attempt to reach Shuster—Freydan—A Georgian colony—Tehrun—Reach Isfahan—M. Boré—Mr. Burgess—The Matamet—The bastinado—Imaum Verdi Beg—Shefi'a Khan—Ali Naghi Khan—Invitation to Kala Tul—Delays in departure—Residence at Isfahan—Messrs. Flandin and Coste—The Palaces—Persian orgies—The Mujtehed 277

CHAPTER VIII.

Departure from Isfahan—My travelling-companions—The Shutur-bashi—Shefi'a Khan—False alarm—Enter the Bakhtiyari country—Fellaut—Hospitable reception—Chilaga—A foray—Lurdagon—A Bakhtiyari feast—Effect of poetry—Difficult mountain pass—Thieves—Reach the Karun—Kala Tul—The guest-room—Mehemet Taki Khan's brothers—His wife—His sick son—The great Bakhtiyari chief—Cure his son—Khatun-jan Khanum—Khanumi—Fatima—Hussein Kuli—Ali Naghi Khan's wives—Dress of Bakhtiyari women—Marriages—Life at Kala Tul—The Bakhtiyari 335

CHAPTER IX.

Excursion to Mal-Emir—Bakhtiyari graves—The Atabegs—A wife of Mehemet Taki Khan—Plain of Mal-Emir—Mulla Mohammed—Sculptures and inscriptions of Shikefti-Salman—Leave Mal-Emir for Sûsan—Robbed on the road—An

Iliyat encampment—Difficulties in crossing the Karun—
Mulla Feraj—The tomb of Daniel—A fanatic—Suspicions
of the Bakhtiyarı—The ruins—Ancient bridge—Bakhtiyari
music — Leave Sûsan — Forest encampment — Return to
Kala Tul—Recover my property—Visit ruins of Manjanik
—Legend relating to Abraham—Ill of fever—Village of
Abu'l Abbas—Attempt to visit Shefi'a Khan—Dangers of
the road—Return to Kala Tul—Accompany Shefi'a Khan to
his tents—A terrible night—Encounter with a lion—The
ons of Khuzistan—Leopards and bears—Recalled to Kala
Tul—Escape from drowning 398

CHAPTER X.

Demands upon Mehemet Taki Khan—He is declared in rebellion—Threatened invasion of his mountains—Requests
me to go to Karak—The trade of Khuzistan—Leave for
Karak—The Kuhghelu—Ram Hormuz—The Bahmei—
Behbahan—Bender Dilum—Mirza Koma—Embark for
Karak—Arrive there—Return to Kala Tul—March with
Mirza Koma—Danger from Arabs—Reach the castle—
Mehemet Taki Khan at Mal-Emir—Adventure with Baron
de Bode—Join Mehemet Taki Khan—Effect of poetry on
Bakhtiyari 449

ILLUSTRATIONS

IN VOL. I.

PORTRAIT OF THE AUTHOR IN THE BAKHTIYARI
    COSTUME. *Taken at Constantinople in 1843.*
    Coloured plate . . . . . . *Frontispiece*

VIEW   THE RESIDENCE OF THE
    BAKHTIYARI CHIEF . . . . . . *to face p.* 364

MAP OF SYRIA . . . . *at end*

MAP OF KHUZISTAN . . . . . . *at end*

EARLY ADVENTURES.

CHAPTER I.

INTRODUCTORY.

I HAVE often been asked how it came to pass that my attention was directed to the ruins of Nineveh, and that I was able to carry on the excavations among them that led to the discovery of the Assyrian remains with which my name has been associated. It is, therefore, possible that an answer to the question may have some interest, and if one is to be given, it should not be much longer delayed.

After my return to England from the East, I occupied an occasional leisure hour in transcribing and putting together in the form of a narrative the notes that I had made during an adventurous journey and residence among some of the wildest tribes in Persia, in the years 1840, 1841, and 1842. I did this for the benefit of those who might hereafter take an interest in my early history, and in such a form that I might, if I thought it desirable, continue my narrative in the form of an autobiography. I lately showed a part of it to a friend, who had often put to me the question to which I have referred, and he has advised me to give it to the press. I have followed his advice, but whether or not I have been judicious in doing so, remains to be seen. I have had neither time nor inclination to rewrite it, and I publish it in its original rough form, without attempting to add to it those reflections and descriptions which are usually introduced by travellers into the relation of their adventures, after they return to their desks. During the exciting and perilous days that I passed in the mountains of Luristan and among the Arab tribes of Turkish Arabia, I had little time for writing an elaborate journal. It was, indeed, as much as I could do to make a few hasty notes, and, living as I was among barbarous people, who entertained the greatest suspicion of my motives for visiting their country, it was often very dangerous to be seen putting pen or pencil to paper. Fortunately, when among the Bakhtiyari, notoriously the most lawless of the Persian mountain tribes, I enjoyed the friendship and confidence of Shefi'a Khan, a man of great intelligence and influence, who, being the Vizir, or Prime Minister, of Mehemet Taki Khan, the great Bakhtiyari chief, was intimately acquainted with everything relating to the country inhabited by the Lurs.[1] Being too enlightened to entertain the suspicions to which I have alluded, he was always ready to answer my questions and to give me any information I required as to its condition, geography, and resources. I was indebted for similar information relating to the Arab tribes occupying the plains to the west of the mountain range of Zagros, or Luristan, to one Seyyid[2] Abou'l Hassan, a native of Shuster, who proved a most faithful and useful friend, without whose assistance I should probably not have been able to pass in safety through this dangerous region. He, moreover, kindly wrote down for me in Arabic characters the names of tribes and places, many of which we visited together. To render such names in English letters as they are pronounced by Bakhtiyari mountaineers, who speak a dialect of their own, and by Arabs, who have their peculiar way of pronouncing certain letters, must frequently lead to error unless the proper orthography of them is also obtained.[3]

[1] The mountain range to the west of Isfahan, dividing Central Persia from the plains watered by the Tigris, is now known as Luristan, and its inhabitants, including the Bakhtiyari tribes, as Lurs.

[2] The title of 'Seyyid' is given in Persia to all those who can prove their descent from the Prophet Mahomet or one of the Imaums. The persons who claim this descent are countless.

[3] I have done my best to follow some system in rendering

The information that I was able to collect relating to the geography, political condition, and resources of the Persian province known as Khuzistan—which includes the ancient Susiana and Elymais, and the mountains inhabited by the Bakhtiyari tribes—was embodied in a memoir which I wrote when residing for a short time at Baghdad in the years 1841 and 1842. It was subsequently published in the 'Journal of the Royal Geographical Society' for 1846 (vol. xvi.). It was written when I was suffering from constant attacks of intermittent fever, and from the depression of spirits and mental lassitude consequent thereon. It could not, therefore, claim any literary merit; but it was the fullest account of a country of great interest, and very little known, which had then—or which, indeed, has since—been published, and it obtained for me the honour of the gold medal of the Society.[4]

Having given in this memoir all the informa-

Turkish, Persian, and Arabic names into English, but I may have failed to do so in many instances. I have generally written, as nearly as I am able, names in common use as they were pronounced.

[4] In addition to the description of the province of Khuzistan, which was published in the *Journal of the Royal Geographical Society* for 1846, I sent to the Society in 1839 a paper on the river Rhyndacus, which flows through a part of the ancient province of Bithynia. I had followed its course, then not fully known to geographers, and incorrectly laid down in our maps. Some further notes on Khuzistan which I sent to the Society were published in vol. xii. of its *Journal*.

tion which I had collected, and which I believed might have any political or scientific interest and might be of any practical use, I confined myself in my narrative almost exclusively to my personal adventures. I must, therefore, ask the indulgence of my readers—should I be fortunate enough to have any—for the too frequent use of the pronoun 'I.' In a narrative of the nature of an autobiography one can scarcely avoid being egotistical.

In the following pages I have described the Persians and Arabs such as I found them. The character that I have given of them will not be considered a favourable one. It is possible that since the time of my residence among them—now more than forty-five years ago—a change may have taken place, and that the misgovernment, oppression, and cruelty which I have denounced, especially in the rulers of Persia, and the vices of all classes of the people which shocked me so greatly, are no longer what they were. I trust that such may be the case, and that I may not be giving offence to any modern Persian, more highly educated and better acquainted with the institutions and customs of civilised nations than his forefathers. No one who has any knowledge of what Persia and its government were half a century ago would probably question the truth of what I have written. Great changes, in many respects for the better, have taken place in Turkey during the last fifty years, although I doubt whether it is as interesting and pleasant a country to travel in as it formerly was; but from all I have read and heard I much fear that there has been little improvement in the administration of Persia, or in the lawless and turbulent habits of her mountain tribes.

Much of my boyhood was passed in Italy, where I acquired a taste for the fine arts, and as much knowledge of them as a child could obtain who was constantly in the society of artists and connoisseurs. I also imbibed that love of travel which has remained to me through life. When about sixteen years of age I was sent to London to study the law, for which I was destined. But, after spending nearly six years in the office of a solicitor and in the chambers of an eminent conveyancer, I determined for various reasons to leave England and to seek a career elsewhere. A relation, who held a high official position in Ceylon, led me to hope that I could find an opening there at the Bar or in the Civil Service, of which I might avail myself with every prospect of success. I resolved, therefore, to go to that island. It happened that my relative was acquainted with another young Englishman who had formed a similar resolution, Mr. Edward Ledwich Mitford, to whom he introduced me. This gentleman, who was some ten years older than myself, and who had a dread of a voyage by sea, proposed that we should perform the journey together, and that it should be as far as possible by land. My love of travel and adventure induced me to accede readily to his proposal. We accordingly agreed to proceed through Central Europe, Dalmatia, Montenegro, Albania, and Bulgaria to Constantinople. Thence to cross Asia Minor to Syria and Palestine, and the Mesopotamian desert to Baghdad, which was to be one of the stages of our long journey. From Baghdad we believed that we should be able to reach India through Persia and Afghanistan, and ultimately Colombo, always travelling by land, except when passing over the narrow strait of Adam's Bridge.

The idea of visiting Aleppo, Damascus, Baghdad and Isfahan greatly excited my imagination, which had been inflamed with the desire to see those renowned cities of the East when as a boy I used to pore over the 'Arabian Nights.' In addition to this fascinating book I had greedily read every volume of Eastern travel that had fallen in my way. I had made acquaintance when in London with Baillie Fraser, whose novels descriptive of Persian life I had devoured with the greatest eagerness, and with Sir Charles Fellowes, whose account of his discoveries among the ruined cities of Asia Minor I had listened to with the liveliest interest, and with an ardent desire to follow in his footsteps. The works of Morier, Malcolm, Rich, and other travellers had given me a longing to visit Persia, Babylonia, and the wild tribes of Kurdistan. Having a vague notion that I might some day be able to see those countries, I had even attempted to master the Arabic characters and to learn a little of the Persian language. Nothing, therefore, could have been more delightful to me than the prospect before me.

As neither my companion's means nor my own would allow us to incur any considerable expense in performing our long journey, we determined to travel with the utmost economy. We thought that we should be able to dispense with a servant, to buy the horses we might require when in the East, where no other means of transport existed, and to find our way with the aid of the compass, having only occasionally recourse to a guide. The plan was a vast and a somewhat romantic and extravagant one in those days, nearly half a century ago, when some of the countries which we proposed to traverse were very little known, and in which a European could scarcely show himself without running considerable risk. But dangers, considerations of health, and the prospect of great fatigue and privations in our 'land-march,' did not enter into our calculations. I was too young, enterprising, and robust in constitution to think of such things, and Mr. Mitford, who was an old traveller in Morocco, made light of them.

I obtained an introduction to Sir John MacNeill, who, in consequence of a misunderstanding between the English and Persian Governments, had recently been withdrawn from Teheran, where he had represented England at the court of the Shah, and who was then in London. We submitted our plans to him and asked his opinion as to their practicability. He encouraged us to persevere in them, thinking that in the then state of affairs in the East, when Russia was suspected of entertaining ambitious designs with respect to the Turcoman Khanats, and of endeavouring to draw Persia away from the influence of England, we might, during our journey, obtain information of use to the British Government, and of value for the elucidation of the geography of a little-known part of Asia. I remember asking him how he would advise us to travel when passing through Persia and Afghanistan. His answer was, 'You must either travel as important personages, with a retinue of servants and an adequate escort, or alone, as poor men, with nothing to excite the cupidity of the people amongst whom you will have to mix. If you cannot afford to adopt the first course, you must take the latter'—and the latter we determined to take.

We also placed ourselves in communication with the Royal Geographical Society, offering our services in clearing up any doubtful geographical questions connected with the countries in Asia through which we intended to pass. It was suggested to us by the Council that instead of taking the route to Herat and Afghanistan through the north of Persia, which had been followed and described by more than one traveller, we should endeavour, if possible, to reach Kandahar from Isfahan through Yezd and the Seistan, exploring on our way the Lake of Furrah. That part of Central Asia had not then been visited, and much curiosity and interest had been excited by the reports concerning it which had reached the Society through native sources. Ruins of ancient cities and of remarkable monuments were said to exist on an island in the Lake of Furrah, and in the country watered by the river Helmund. But all attempts to reach them had hitherto failed, and the life of any European who ventured into the Seistan would, it was believed, be in great danger. We promised to take this route if, on reaching Isfahan, we found it practicable to do so.[5]

[5] Some time after we had left England Dr. Forbes was murdered in an attempt to visit the Seistan.

My attention was also called to a paper on Susiana by Sir Henry, then Major, Rawlinson, which had been recently published in the 'Journal of the Geographical Society.'[6] This distinguished Oriental scholar was already known for his researches and discoveries connected with the ancient geography and languages of Persia. When an officer in the East India Company's service, authorised to serve in the Persian army, he had held command in a military expedition sent against the rebellious populations of Khuzistan. Under these favourable conditions he was able to examine a country in which an ordinary traveller, depending upon his own resources, would have run very great danger.[7] Although with the protection he enjoyed he visited several important ruins, identified the sites of many ancient cities, and cleared up many disputed points in the geography of Susiana and other parts of Western Persia, he was unable to visit the highlands inhabited by the Bakhtiyari tribes, who were always more or less in open rebellion to the Shah. He had, however, opportunities of communicating with some of their chiefs, and received from them descriptions of ruins and rock-cut inscriptions

---

[6] 'Notes on a March from Zoháb to Khuzistan,' *Journal of the Royal Geographical Society*, vol. ix.

[7] Captain Grant and Lieutenant Fotheringham, two officers in the Indian army, were murdered when attempting to explore it some years before.

which they alleged existed in their mountains. The Royal Geographical Society considered it very desirable that these ancient remains—and especially the ruins known by the Lurs as Sûsan, which Major Rawlinson was induced to believe represented the site of 'Shushan the Palace,' where the Prophet Daniel saw the vision[8]— should be thoroughly examined. I promised, therefore, to visit, if possible, the Bakhtiyari Mountains, thinking that I might be able to do so on my way from Baghdad to Isfahan. I carried a copy of Major Rawlinson's highly interesting memoir with me, and it served me as a text-book during my travels and sojourn in the country to which it relates.

I was desirous, in order to make my journey as profitable as possible, to learn the use of some simple instruments for taking observations, for finding the latitudes of places whose exact positions had not been determined, and for ascertaining the heights of mountains, and at the same time something of surveying. This would enable me to map out my route, if not with scientific accuracy, at least roughly. I was introduced by a friend to an old retired captain of the merchant service who lived in the city. He undertook, for a very small remuneration, to give me lessons in the use of the sextant, and taught me how to take observations

[8] Daniel viii.

of the sun for the latitude, and to fix the positions of mountain peaks and other objects, such as towns and ruins, with reference to my route. The instruction which he gave me was of the most elementary kind; but it enabled me to add not inconsiderably to the maps of the countries which I traversed. I provided myself with a pocket sextant, an artificial horizon, a Schmalcalder or Kater's prismatic compass, a telescope, some thermometers for determining the temperature and ascertaining heights, an aneroid barometer, and a silver watch. By the advice of Sir Charles Fellowes, who had recently returned from his explorations in Asia Minor, I had the watch painted black, so that the sight of the bright metal might not excite the cupidity of the wild people whom I should encounter, and who would not hesitate to rob and even murder a traveller to obtain possession of any object supposed to be of value. Most of these instruments were stolen, broken, or lost in the course of my travels.

I was further desirous of learning something about medicine and the treatment of common diseases. A medical man with whom I was well acquainted kindly offered to give me such instruction on the subject, within the short space of time at my disposal, as he thought might be useful to me. He took me to the London University

Hospital, explained to me the symptoms and treatment of the diseases, such as intermittent and other fevers, ophthalmia, and dysentery, that I was most likely to meet with, taught me the use of the lancet, how to deal with simple wounds, and how to stop the bleeding of an artery by a tourniquet. He further provided me with the most necessary medicines, wrote me out instructions as to the use of each, and supplied me with some vaccine lymph. The directions that I received from him proved of the greatest use to me during the course of my travels, both in treating myself when suffering from fever and other maladies, such as diarrhœa and dysentery, and in prescribing for the patients who came to me under the conviction that all Europeans were skilful 'hakims,' and could cure every imaginable disease.

My companion had some knowledge of natural history, especially ornithology, and botany. He provided himself with instruments required for skinning and stuffing, and as we were each to carry a double-barrelled gun, to serve for defence as well as to provide us with game for food in case of necessity, we looked forward to making a collection of birds, as well as other collections, on our way. We soon, however, found that it was impossible to do so to any extent, as we possessed no means of transport, our baggage having been reduced to the smallest possible compass, not exceeding what we could carry in our saddle-bags. In addition to a little linen and a change of clothes, I provided myself with what was called a 'Levinge bed'—a pair of sheets sewn together and attached to a mosquito curtain, forming a kind of bag, which, when closed (the curtain being attached to a nail in the wall), formed a complete defence against insects of all kinds, whether crawling, hopping, or flying, that abound in the dirty houses and still more filthy caravanserais that we were warned we should have to occupy when travelling in the manner we intended to do in the East. This 'Levinge bed' proved of the greatest comfort to me, and insured me many a refreshing night's rest, after a long and fatiguing day's journey, in stables and other places swarming with vermin. It took up little room in my saddle-bags, and I clung to it as long as I possibly could. As it did not offer any temptation to those who more than once relieved me of the greater part of my little property, I was able to retain it during a considerable part of my wanderings in the East. A long cloth riding-cloak served as some protection from rain, and when sleeping on the bare ground. It could be rolled up and strapped behind my saddle. I procured a letter of credit for 300*l*. from Messrs. Coutts, which, owing to some mistake on the part of one of their agents, was very nearly proving the source of very serious trouble and vexation to me, as two of my drafts given to gentlemen who had kindly advanced me money in the East were returned dishonoured, and, in consequence of there being no means of communicating with me, remained long unpaid.

At length all our preparations were completed, and on July 8, 1839, I left England by way of Ostend for Brussels, where I was to be joined by Mr. Mitford, who had preferred the shorter sea route by Calais. As he has published a narrative of our journey together as far as Hamadan, in Persia, without, however, mentioning my name as that of his companion,[9] I shall not in the following pages repeat the description he has given in his work of our travels, but shall confine myself to relating the adventures that befell me when I left him for a time at Jerusalem to visit Petra and other sites in the Syrian desert, and to the story of my life in Persia and the Bakhtiyari Mountains, and among the Arab tribes, after we finally separated at Hamadan.[1] I need only say that on our

[9] *A Land-March from England to Ceylon Forty Years ago, through Dalmatia, Montenegro, Turkey, Asia Minor, Syria, Palestine, Assyria, Persia, Afghanistan, Scinde and India,* by Edward Ledwich Mitford, F.R.G.S. 2 vols. London. W. H. Allen & Co., 1884.

[1] Mr. Mitford states in his work (vol. i. p. 369) ' my companion,

way to Constantinople through Rumelia I caught a fever, by sleeping, I believe, at a posthouse in a marshy plain near Philippopolis. At Constantinople it developed itself into a dangerous gastric attack, which confined me for some time to my bed. I was attended by Dr. Z., an Armenian gentleman who had studied medicine at Edinburgh. He bled me twice copiously, and, moreover, made a large circle with a pen and ink on my stomach, which he ordered to be filled with leeches. My strength was so much reduced by this great loss of blood that I was unable to accompany my fellow-traveller on his journey by land round the Gulf of Ismid or Nicomedia, and through Bithynia to Mudania, on the gulf of that name. I joined him at the latter place, having taken passage on board a bazar caïque, or public boat employed in transporting passengers from Constantinople.

On January 9 we reached Jerusalem. The

finding that we should not be able to follow the route we had intended, resolved to return to Bushire, on the Persian Gulf, while I prosecuted my journey alone through Khorassan, Afghanistan, and India.' But I left Mr. Mitford because I was then determined, as it will be seen in the course of my narrative, to persist, if possible, in our original intention of making our way through Yezd and the Seistan to Kandahar. The events which in the end caused me to abandon my intention will be related. Mr. Mitford further mentions (vol. i. p. 218) my having left him at Jerusalem 'for an excursion in the Hauran:' My object was to visit Petra, Ammon, Gerash, and other ruins of ancient cities in the Syrian desert.

intense cold we experienced there, and how our travels and our lives were very nearly brought to an untimely end by the fumes of a charcoal brazier which we had imprudently kept in the room in which we slept, has been related by Mr. Mitford.

The descriptions that I had read of Petra and of the remains of ancient cities to the east of the Jordan in the works of Burckhardt, Laborde and the few other travellers who had then been able to reach them, had given me an intense longing to see them. I proposed to my fellow-traveller to endeavour to reach Damascus through the desert and the Hauran instead of by the usual route, visiting these ruins on our way. But he made serious objections to my proposal. We were, he pleaded, in mid-winter, and the weather, which was unusually cold and rainy, was not favourable to an excursion in which we should be without shelter. We had, moreover, been assured that it would be impossible to pass through the dangerous country beyond the Dead Sea with safety without the protection of some powerful Arab sheikh and a strong escort, for both of which we should have to pay a considerable sum of money, and with our limited means this we could not afford to do.

My companion's objections to my proposal were no doubt well founded, and had my experience been greater and had I been less headstrong, they would have prevailed with me. But I was resolved not to be deterred from my scheme, and as he declined to accompany me I determined to undertake the journey alone, trusting to my own resources, and fully believing all the romantic stories that I had read of Arab hospitality and their respect for a guest.

The English Consul, Mr. Young, in vain attempted to dissuade me from my project, which he denounced as foolhardy and impracticable. Finding that I was obstinate, he gave me to understand that he could in no way be held responsible for anything that might happen to me. He then offered to give me any assistance in his power, and kindly procured me letters from the Egyptian authorities[2] to persons of influence at Hebron, who, he believed, might obtain for me the protection of the sheikhs of the Arab tribes through whose territories I should have to pass. He, however, warned me that those chiefs were known for their rapacity and their lawlessness, that they recognised no authority, and were generally at war with each other, and that if I escaped being murdered I should in all probability be robbed of

[2] It will be remembered that at that time Syria and Palestine were in the possession of the Egyptians under Ibrahim Pasha, the son of Mehemet Ali, who had recently defeated the Turkish army at Nizib.

everything I possessed, and left stripped to the skin to find my way back to Jerusalem.

My companion, who wished to spend a few days more in Jerusalem, proposed to follow the usual caravan route to Damascus, and to wait in that city until I rejoined him.

Although I had picked up sufficient Arabic to suffice for ordinary purposes, I thought that it would be advisable to take an interpreter with me. I could not find one who was disposed to encounter the dangers of the journey, but I fell in with an Arab youth who came originally from the country to the east of the Dead Sea, had been converted to Christianity, and spoke that mongrel Italian known as the 'Lingua Franca,' which he had picked up from the friars in whose service he had been. He had received on conversion the name of Antonio, by which he was known. He offered to accompany me, and as he appeared active and intelligent I engaged him as dragoman and servant.

CHAPTER II.

Leave Jerusalem — Bethlehem — Latins and Greeks — Hebron — Colonel Yusuf Effendi—The bastinado—Sheikh Abu-Dhaouk —Departure for Petra — The tents of Abu-Dhaouk — Savage country—Danger from Arab robbers—Meet Bedouin horsemen —A sacred spot—Enter the mountains—The Wady Musa— Ascend a peak—Rock-cut monuments—Reach Petra—Difficulties with the Arabs—The ruins.

ON January 15, 1840, I left Jerusalem for Petra, accompanied by the Arab boy Antonio. I had hired two mules to carry us as far as Hebron, where I hoped to make arrangements to continue my journey in the desert. I purchased a small bell tent which belonged to an Egyptian soldier. It was old and worn, but would serve to give me shelter at night, and could be easily pitched. I took with me a little store of rice and flour for food. These things, with my carpet, my saddle-bags containing a change of linen, maps and note-books, a compass and a small supply of medicines, constituted the whole of my baggage. I carried my double-barrelled gun for defence, and, at the same time, in the hope that it might enable me to provide myself with game when meat was not to be obtained.

On my way to Hebron I stopped at Bethlehem. I was much struck by the picturesque aspect of the place, approaching it from the north. The Convent and Church of the Nativity rose boldly on the summit of a hill clothed with olive and fig trees, which contrasted with the barren and desolate uplands which I had crossed after leaving Jerusalem. I remarked the dress of the women who were at the wells, or whom I met in the streets as I entered the town. It was that which the old painters had given to the Virgin—loose drapery of blue and red, with a white kerchief thrown over the head, the ends falling on the shoulders. Many wore silver coins strung together as ornaments round their foreheads, and, as in other parts of Palestine, I observed among the girls no little beauty and much grace and elegance of form.

I rode to a convent to which I had been directed, but it was some time before I could obtain admittance. At length the iron-bound door, about four feet high, was unlocked and unbarred, and I crept in. The place had the appearance of a fortress, whose inmates were in a state of siege and feared an enemy. Such was, in fact, almost the case. But it was not the Turks or Arabs that the inhabitants had to fear, but their Christian brethren of the orthodox Greek faith. They were 'Latins,' or Roman Catholics, and a feud, the more bitter and irreconcilable because it was religious, existed between them and the rival community. It had not long before led to a pitched battle, in which the Greeks had remained the victors, and had ended by ejecting the Latins from the Church of the Nativity. The scandalous scenes which had occurred between the two sects at Jerusalem had been repeated at Bethlehem, and the Catholics now feared that they might be turned out of their convent by their persecutors. Hence the precautions that were taken before even a solitary stranger was admitted. I was received by a Franciscan friar, who, after carefully closing and barring the door, conducted me into the refectory. There I was courteously welcomed by the superior and several brothers. They were at the early dinner, and invited me to join them.

After I had eaten I was taken to the church and was shown the grotto in which the Saviour was born, the manger in which He was laid, the burial-place of the Innocents slain by Herod, and the other sacred spots which the priests and monks have successfully identified for the edification of the faithful and for the devotion of pious pilgrims, as well in their own interest as in that of the religious establishments to which they belong.

I had to listen to the complaints of the Latins against the heretical and impious Greeks who had endeavoured to possess themselves of all these holy places, and had even sought to exclude their rivals by a wall of masonry from the high altar, thus defiling and disfiguring the Church of St. Helena, and compelling the Roman Catholic monks to celebrate their Mass in a small and inconvenient side-chapel, although they boasted a congregation of about two thousand of the faithful. The very spot where Christ was born had been monopolised by these detestable heretics, under the protection and with the support of Russia, and they had even carried their impious audacity so far as to profane the Holy Place by inserting a silver star in the pavement However, my guide boasted that the Latins had succeeded in keeping possession, notwithstanding the violence and outrages to which they were exposed, of the very manger in which the Child was placed, and which had very much the appearance of one of those troughs for the reception of the dead found in tombs excavated in the rock, of which I had seen large numbers in Syria and Asia Minor. The caves covered by the Church of Bethlehem resemble indeed so closely such excavations, that had a doubt been permitted as to the truth of the traditions which attach to them or which have been invented concerning them, they might also be taken for the sepulchres of the early inhabitants of the land.

After I had thus performed my pilgrimage I quitted Bethlehem, as little edified as I had been at Jerusalem by the animosities and hatreds of rival Christian sects. The road again wound through barren and solitary hills, with here and there the ruins of a deserted village. Some interest was, however, imparted to my journey by occasional views of the distant mountains of Moab, and glimpses of the blue waters of the Dead Sea. I stopped for a short time at the so-called 'Pools of Solomon,' which have been identified with the reservoirs constructed by Hezekiah for supplying Jerusalem with water by an aqueduct, of which traces still remain.

Hebron was so well concealed in a small valley that I found myself in its streets before I knew that I had reached the end of my day's journey. The hills about it were clothed with vineyards, yielding fruit renowned in Palestine. The vines were dressed in a peculiar and, to me, unusual fashion. They had the appearance of standard rose trees, the lower leaves and branches being lopped off and the stem being only allowed to attain the height of about five feet, with a tufted head. From the grapes which they produced the Arabs made a kind of molasses, called 'dibs.' A few olive and pomegranate trees grew here and there, but there appeared to be a general absence of wood. The place itself was in a ruined state. I observed in the neighbourhood numerous tombs excavated in the rocks, a mode of sepulture evidently at one time generally practised in this as in other parts of Syria.

I had a letter for a native Christian, of the name of Elias, who, I had been informed, was in the habit of entertaining travellers. After some trouble I succeeded in finding his house, but only to learn that its owner, who had been a collector of taxes for the Government at Hebron, had been thrown into prison on account of some alleged irregularity in his accounts. Whilst inquiring my way to the house of the Muteselim, or governor, to whom strangers were then in the habit of applying for a night's lodging, I met in the street a Turk named Yusuf Effendi, a colonel in the Egyptian army, of whom I had heard at Jerusalem. His friends there had proposed to give me a letter of introduction to him, but they had not done so as they believed him to be absent from Hebron. Although not the governor he was higher in authority. I ventured to address him, and placed the letter to the Muteselim in his hands. He received me with the polite courtesy of a well-bred Turk, and at once invited me to take up my quarters with him. The house in which he lived was half in ruins, like most of the others in the town, but he could offer me a room in it.

The Muteselim arrived soon after on a visit to the colonel, whose room was soon crowded with persons coming to pay their respects or having business to transact. In the middle of it an old hag, having much the appearance of a witch, her long grey hair falling over her tattered garments, was preparing a potion of herbs over a brazier. She had the reputation of being skilled in medicine, and Yusuf Effendi, who was suffering from a cough, had sent for her. After she had sufficiently boiled her draught, she pronounced some mystical words over it and presented it to her patient, who drank it off, whilst she felt his pulse and muttered some sentences which were to act as a further charm. She then retired. All this was done in public, the room being thronged with people. The colonel, who had the reputation of being an enlightened and educated man, seemed to have no misgivings as to the power of the old woman to cure him, and of the efficacy of her remedies.

Yusuf Effendi gave me an excellent supper after the Turkish fashion, some fifteen dishes having been placed one after the other on a metal tray, supported upon a low stool, and removed so quickly that we had scarcely time to taste of them. We sat on the ground and ate with our fingers, which we wiped upon large flat cakes of bread. In the evening my host received numerous visitors, amongst whom was Abdua'l-Jewâd, the principal sheikh and mufti of Hebron, for whom I had also a letter. My proposed journey to Petra was discussed and the dangers and difficulties of it insisted upon; but, as I persisted in my determination to attempt it, the sheikh and the governor promised to give me such assistance as they were able.

The district of Hebron and the Arab tribes on its borders had been recently in rebellion against the Egyptians. The inhabitants of the town had risen and had expelled the governor, who had taken refuge in Jerusalem. Yusuf Effendi had been sent with some troops to re-establish his authority and to punish the rebels. He had occupied the town and had just returned from an expedition against the insurgent Arabs, in which he had been completely successful. It was thus that the road to Jerusalem, which a few days before had been closed to caravans and travellers, was now open, and that order had been restored in the country around Hebron. Several of the principal inhabitants of the place, who were accused of resisting the Egyptian authorities, had been thrown into prison. Amongst them was the Christian Elias, for whom I had a letter, who was charged with having taken advantage of the confusion arising from the revolt to help himself from the Treasury chest.

On the morning after my arrival the house occupied by Yusuf Effendi was filled with a crowd of screaming and gesticulating Arabs. Some were the chiefs of villages and of neighbouring tribes who had been summoned by him to account for their recent bad conduct, and to pay arrears of taxes and fines imposed upon them for rebelling against the Government. Others were in chains and had been brought before him to receive their sentences and their punishment. The latter was summary enough. Several stalwart soldiers stood ready with their 'courbashes,' or whips of hippopotamus hide. The culprit who had been convicted, or the accused who could not be brought to confess, was speedily thrown to the ground on his belly and his feet passed through two nooses of cord attached to a stout pole, which, hoisted on the shoulders of two strong men, presented the soles to receive the bastinado. This was inflicted without mercy, water being constantly poured over the wounds to increase the torture, the victim either bellowing loudly or suffering his agony with suppressed groans and cries of Allah! Allah!

The blows, administered by two men who were constantly relieved, fell rapidly upon his bleeding feet. After the number inflicted was deemed sufficient by the colonel the wretched sufferer, unable to walk, was dragged away by the guards.

I felt too much disgusted and horrified with these barbarous proceedings to continue to witness them. Leaving Yusuf Effendi to punish his prisoners, and to extract the fines he had imposed and the taxes in arrears after this fashion, I went in search of the sheikh whom I had met the previous evening, and who had promised to allow me to see as much of the mosques which cover the traditional tombs of Abraham and of the patriarchs as could at that time be shown to a Frank and a Christian unbeliever. Hitherto, I believe, no European had entered the sacred buildings, which were surrounded by a double wall; but one or two travellers had been allowed to see as much as could be distinguished of them from the top of the outer wall. More than one person of distinction has since been admitted into the interior of the mosques. Before the Egyptian occupation, a European and a native Christian even ran some risk in appearing in the town, which was known to Musulmans as 'El Khalil,' or 'The Holy.'

Sheikh Abdua'l-Jewâd, true to his promise, took me to the first enclosure, and to the buildings forming part of it. From windows in them, and from their flat roofs, to which I was permitted to ascend—a privilege then rarely, I believe, granted to Europeans—I was able to look into the inner court or quadrangle in which are the three mosques covering the sepulchres held holy by the faithful. The principal one, into the interior of which I could dimly see, was said to contain the tombs of Abraham and his wife; a second, with two domes, those of Isaac and his wife; and a third, also with two domes, those of Jacob and his wife. These buildings have since been visited by persons competent to deal with the traditions which attach to them, and to determine their age and describe the style of their architecture. The notes that I was able to make during my hasty and imperfect survey are consequently of no value. I find, however, that I came to the conclusion, from some of its architectural features, that the principal mosque had been originally a Byzantine basilica, to which the Musulmans had made additions, such as an exterior portico or arcade.

After I had examined as much as I was allowed to see of the sacred buildings I visited the glass works, in which were then made the bracelets and ankle bangles of blue and red opaque glass worn by the Arab women. Hebron, at that time, had a considerable trade in these articles. They were sold for the small price of from two to five paras (under a penny) apiece.

On my return to the house of my host for breakfast, I found that he had suspended the bastinadoing, and having dismissed the crowd, was enjoying his 'narguilé,' or water-pipe, after his morning's work. He informed me that he had not forgotten my wish to proceed to Petra, and hearing that among the Arab sheikhs who had come to Hebron to settle their accounts with the Egyptian authorities, there was a chief of the Howitat tribe which were encamped in the neighbourhood of the ruins, he had sent for him to make arrangements for my journey. The sheikh soon afterwards appeared. He was a dirty, truculent-looking fellow, with very black eyes and very white teeth, a sinister expression, and complexion scarcely less dark than that of a negro. He offered to hire me two camels to take me to Petra and thence to Kerak, to the east of the Dead Sea; but he declared that the risk and danger were great, for the Arab tribes were at war with each other, and although he could insure my safety amongst his own people, he would not undertake to protect me against all Arabs whom we might meet on the road. Parties of Bedouin horsemen, he said, taking advantage of the unquiet state of the country, were known to be moving about on marauding expeditions. I might fall in with some of them on my way, and as they would probably be his enemies, they would not respect the safe conduct that he could give me. It would consequently be necessary, he maintained, to send an escort of armed men with me for my protection. Even the inhabitants of Kerak, a town to the east of the Dead Sea, were only in nominal subjection to the Egyptian Government, and had refused to receive the troops which the Muteselim of Jerusalem had sent a few days before to occupy the place. Under all these circumstances he could not undertake, he said, to furnish me with two camels and to conduct me to Petra and Kerak for less than two thousand piastres, or 20*l*.

I was persuaded that he had greatly exaggerated the dangers and difficulties of the journey in order to justify this exorbitant demand, to which I peremptorily refused to listen. Yusuf Effendi expressed his indignation at it, and declared that he would send me to both places with a single Egyptian soldier, and would retain the sheikh as a hostage until I had accomplished my journey, holding him personally responsible for my safety. Against this arrangement Sheikh Abu-Dhaouk Haj Defallah, of the tribe of Jehalin—for such, as far as I could make out from his pronunciation of it, was his full name—loudly protested. After a great deal of threatening on the part of the colonel, and the usual guttural vociferation on the part of the Arab, in which the governor of Hebron joined, although engaged, at the time, in his prayers, the sheikh reduced his demand to five hundred piastres, and further undertook to send his own brother with me as a guide, who was to return with a written declaration from the Mujelli, or hereditary chief, of Kerak, that I had been delivered over to him safe and sound. After he had thus come to terms with me, through the intervention of Yusuf Effendi, he left me, promising to call for me with the two camels on the following morning. But he returned in the afternoon, and finding me alone endeavoured to extort from me a further sum by way of 'bakshish.' A Frank, he declared, had recently been asked by another Arab sheikh no less than 40 purses (about 200*l.*) to be taken to Petra, and had been compelled to pay 4,000 piastres. Why should he receive less? However, I refused to depart from our agreement, and threatened to complain to the colonel, upon which he left me. The large sums which had been paid by some European travellers to Arab sheikhs for their protection when visiting Petra and other sites to the east of the Jordan and Dead Sea had excited the cupidity of the Arabs, had rendered them grasping and insolent, and had greatly increased the dangers of a journey in the Syrian desert and Sinaitic peninsula. They no longer exercised the proverbial hospitality of the Bedouin, but had become the most degraded, untrustworthy, and treacherous of their race.

In the morning the two camels appeared with an Arab, but not the sheikh, who sent word that he would join me on the road. Having taken leave of my obliging and hospitable host, I mounted one of the camels, first covering its rude pack-saddle with my carpet. The youth I had engaged at Jerusalem rode the other, which also carried my tent and little baggage. We then left Hebron, accompanied by the camel-driver on foot.

Abu Dhaouk did not join me as he had promised. I soon discovered that the camels did not belong to him, but that he had hired them to take me to his tents. Their owner, who pretended to be acquainted with the road, confessed before long that he did not know where the encampment of the sheikh was to be found. We lost our way, and wandered about for several hours over the barren and stony hills. Not having been accustomed to ride a camel the motion fatigued me greatly. Once, when descending a steep hill, my beast, which I had been in vain urging to quicken its solemn pace, took a fancy to start off in a kind of awkward gallop. I endeavoured to check it with the halter which was fastened to its nose, but it only turned its head round as I pulled with all my might, and looked me full in the face, without stopping or even slackening its speed. I clung to the pack-saddle. My saddle-bags first fell off, my carpet followed, and, losing my balance, I slipped over the tail of the animal and came full length to the ground. Fortunately I was not hurt. An exciting chase then ensued, and it was some time before the driver succeeded in catching his camel. Whether it had taken a sudden fright from something on the road, or whether it resented being ridden by a Frank, ignorant of its habits and ways, I did not ascertain. But it was evident that there must have been some cause for its sudden departure from its usual sober pace. I remounted and had no more cause to complain of similar extravagances.

There was every prospect of our passing the night in the open, as we had lost our way and the country appeared to be a desert without inhabitants. At length, as night was approaching, we met a solitary Arab on foot. He belonged to the tribe of whose tents we were in search, and as he was going to them he offered to be our guide.

During the day I had passed through a few interesting places, or at least places with interesting names, suggestive of high antiquity or of biblical traditions. Palestine abounds in such. At no great distance E.S.E. of Hebron we came to a ruined village with a few inhabitants, called Nebbi Lout— the prophet Lot. A little beyond I saw some ruins named Rieff by my guide, near which were many excavated tombs and chambers in the rocks. The hills we crossed were steep, stony, and barren. The valleys by which they were intersected ran in the direction of the Dead Sea. In one of these valleys, called Wady Salesal, were the black tents of Abu-Dhaouk, which we reached shortly before sunset. The sheikh had not arrived. We were received in his absence by his brother. I pitched my small tent, which was soon filled and surrounded by men of the tribe, who collected to hear the cause of my coming. They had to be satisfied with the explanations which the camel-driver and Antonio, the Arab boy, were able to give them. As the sheikh had promised to furnish me with camels, they would be ready, they said, early in the morning. In the meanwhile I was his guest, and a sheep would be immediately slain for my entertainment.

The flocks were returning from the pastures, and long lines of sheep and camels descended from the hill-tops The sheep were folded in the enclosure formed by the tents, and the lambs allowed access to them, bleating as they discovered their mothers, caused so much noise and confusion that the voices of the Arabs were almost drowned. I was invited to the tent of the sheikh, and soon after I had taken my seat in it his wife appeared, as she said, to welcome me and to present her children to me—four handsome, dirty, and half-naked boys. The real object of her visit was, however, to ask me for some tobacco, which I gave her.

A great mess of rice and boiled mutton was brought to the tent about two hours after my arrival. The sheikh's brother and his friends ate with me, dipping their fingers into the large wooden bowl and picking out the savoury bits, which they presented to me. The night was cold, and I was not sorry to sit before a fire of blazing faggots, until it was time for me to retire to my little tent for the night. I remained for some time at the entrance, gazing on the strange and novel scene before me. It was my first acquaintance with an Arab encampment and Arab life. A full moon in all its brilliancy lighted up the Wady, so that every feature in the landscape could be plainly distinguished. The fires in the Arab tents studded the valley with bright stars. The silence of the night was broken by the lowing of the cattle and the hoarse moanings of the camels, and by the long mournful wail of the jackals, which seemed to be almost in the midst of us.

Sheikh Abu-Dhaouk arrived in the night. He came to me early in the morning, and apologised for not having been at his tents to receive me. He had been detained, he said, at Hebron, by the governor, who insisted upon the payment of a much larger sum as tribute from the tribe than the Government was entitled to, and had threatened to throw him into prison unless it were forthcoming. I had slept little, as I was suffering greatly from a toothache. The sheikh declared that there was a skilful dentist in the encampment; and as the pain was almost unbearable, I made up my mind to put myself in his hands rather than endure it any longer. He was accordingly sent for. He was a tall, muscular Arab. His instruments consisted of a short knife or razor, and a kind of iron awl. He bade me sit on the ground, and then took my head firmly between his knees. After cutting away the gums he applied the awl to the roots of the tooth, and, striking the other end of it with all his might, expected to see the tooth fly into the air. But it was a double one, and not to be removed by such means from the jaw. The awl slipped and made a severe wound in my palate. He insisted upon a second trial, declaring that he could not but succeed. But the only result was that he broke off a large piece of the tooth, and I had suffered sufficient agony to decline a third experiment.

After I had undergone this very disagreeable and unsuccessful operation, the sheikh, whilst expressing his sympathy for me, suggested that the sum he was to receive for the hire of his camels and for his protection was not sufficient, and that the agreement which he had made with me had been extorted from him by Colonel Yusuf Effendi. If, therefore, I insisted upon his adhering to it, I ought to make a voluntary addition to it by way of bakshish. I refused to do so, and as he began to make difficulties about finding camels and a man who was willing to run the risk of accompanying me, I threatened to return at once to Hebron and to refer the matter to the colonel. Finding that I was resolute and was preparing my baggage, he gave way somewhat sulkily. But it was already ten o'clock before the two camels were forthcoming. Instead of sending his brother with me, as he had promised to do, he brought two Arabs on foot, armed with long guns, who, he said, would accompany me as guards as well as guides, as the country was very unsafe. I was under the necessity of yielding, and at length, after many delays and much squabbling, I left his encampment.

Our course was due east, and when we reached the summit of the hill which overlooked the tents of the sheikh, we came in view of the Dead Sea. As I stopped to gaze at it a party of horsemen, armed with long spears tufted with ostrich feathers, rode up to me. They looked as if they were returning from a marauding expedition. They belonged to the tribe of Abu-Dhaouk, and after saluting me and embracing my two guards, they asked some questions concerning myself and the object of my journey, and then went on their way. As we continued winding slowly over the barren and stony hills which still separated us from the Dead Sea, we frequently passed Bedouin horsemen, or saw them in the distance. Antonio, who had become a coward by having been brought up in a convent by monks, was persuaded that every Arab we saw, far and near, was a robber. But, whether he was right or not, we were not molested. After a very toilsome and weary journey, owing to the slow pace of our beasts, we suddenly discovered, in a sheltered valley, a number of Arabs who were on their way to the encampment we had left in the morning, with a large caravan of camels laden with corn. They had stopped for the night, and their camels, released from their loads, were kneeling in a circle around them. I pitched my tent near them, and they went in search of water for me.

They returned with some, which was so thick and muddy that, although suffering from thirst, as we had met with no water during the day, I could scarcely bring myself to drink it or to use it for cooking my rice.

My two guards, whose names were Musa (Moses) and Awad, declared that the country through which we had to pass was so dangerous, owing to Arabs who, flying from the military conscription, had concealed themselves in these wild and desolate hills, and lived by robbing travellers, that it was safer to travel through it by night. By three o'clock in the morning the camels were loaded, and we were again on our way by moonlight. Descending rapidly by a very rough and precipitous path, down which we had to lead the camels, we found ourselves in a wild and weird glen, surrounded by detached rocks of sandstone of the most fantastic shapes. The summit of one of them was crowned with the ruins of a small castle, and at its foot were the remains of an ancient building of considerable size. Hard by was an artificial reservoir, in which there was a supply of rain-water. The Arabs called the place Kalat-ez-Zoer, and the name suggested that the site might be identified with that of the biblical Zoar. A more savage and desolate spot I had never seen, and the dark shadows in the moonlight added to its somewhat awful aspect. It was a fit haunt for robbers.

We continued to wind through this inhospitable region, following a narrow valley formed by precipitous rocks. The solitude was undisturbed by the noiseless tread of our camels. The guards enjoined strict silence, as the slightest noise might betray our presence to the robbers who, they were persuaded, were lurking in the caves. They even went so far as to vow to sacrifice a sheep, and give its flesh to the poor, if they passed safely through the dangerous defile into which we had entered. Whether in consequence or not of this pious vow, we emerged safely from it in about an hour, and found ourselves among low sandhills about a mile from the Dead Sea. My guides proposed that we should stop to breakfast in a sheltered nook in which we could conceal ourselves.

After we had taken some rest we resumed our journey, avoiding the shores of the Dead Sea, and following a narrow gully in places scarcely ten feet in width, formed by natural walls of sandstone from thirty to three hundred feet in height, and curiously and intricately stratified. This was the lowest of the parallel ranges of hills which form as it were gigantic steps and terraces from the heights of Jerusalem to the level of the Dead Sea—treeless, waterless, barren, and desolate.

We continued for about an hour in this gully, and then, crossing the low hills, or rather mounds, of sand which formed its eastern side, descended to a salt marsh which appeared to have been left by the receding waters of the Dead Sea. Passing some similar sandheaps we came to a spring of brackish, ill-flavoured water, in which I found a number of small shells. As there was no other on our road for some hours, we were under the necessity of filling our skins from this spring—which the Arabs called the 'Ain Arous'—the eye of the bride. Soon after leaving it we met a poor Arab boy who was wandering in this desolate region in search of the tents of his tribe, and had not, he declared, tasted food for thirty hours. We gave him some bread, and directed him on his way.

We now struck into a deep ravine running due south and away from the Dead Sea, and formed by the same barren sandhills through which we had been wandering during the morning. It was four o'clock in the afternoon before the guides found a retired spot with a little water in which they thought we could encamp in safety for the night. The camels were unloaded and turned loose to pick up such provender as they could find in the coarse and scanty herbage of this desolate region. They had no other food, and as they journeyed along stopped at almost every step to munch a thorny plant which grew in the sand. This, added to their disagreeable and wearying motion, rendered it very trying to ride them. It was useless to urge them onwards. They possessed the calmest and most imperturbable of tempers, and treated blows and imprecations with equal indifference.

The earth in the neighbourhood of the Dead Sea is so impregnated with salt and bituminous matter that even the pools of water formed by recent rains become speedily brackish and noisome to the taste We appeared to be in the bed of an ancient river or torrent running in the direction of the Gulf of Akaba. Huge boulders lay in all directions, and amongst them the semi-fossilised trunks of palms and other trees, which must have been washed down from the valleys and heights to the east of the Dead Sea.[1]

Although we had selected a retired spot in which to conceal ourselves, the guards considered it necessary that incessant watch should be kept during the remainder of the afternoon, and after dark, in case we had been observed by robbers on

---

[1] The physical nature and aspect of this singular and interesting region have of late years been so fully examined and described that the few notes I made on the subject would be now valueless.

the look-out for travellers. They called me at two o'clock in the morning, and insisted that I should take advantage of the moonlight to continue my journey through this dangerous district. Leaving the narrow valley, we emerged upon a sandy plain bounded to the east by a range of high mountains. serrated and fantastic in their outlines. I was allowed to stop at eight o'clock to get some breakfast, but was soon hurried on again. We had scarcely started when Musa, who had been acting as a scout, returned in much alarm, declaring that he had seen horsemen in the distance. From their appearance they could only, he said, be robbers, and he recommended me to prepare for an attack. I was disposed to believe that he had been deceived, or that his fears were exaggerated, when I observed three Bedouins on horseback, with their long tufted spears, coming towards us from different directions. It was evident that they had been watching us, and, from the manner of their approach, that they had evil intentions. Musa made a counter-movement and took up a position to the right of the advancing Arabs, and, levelling his long gun at the foremost of them, prepared for defence. Seeing that there was apparently cause for the alarm of my guides, I threw off my cloak, and, slipping over the tail of my camel to the ground, prepared for action with my double-barrelled gun, which was loaded with ball. Awad, the other guide, also made ready to defend himself and the camels.

The robbers—for such they now proved themselves to be—seeing that we possessed firearms, whilst they only had spears, commenced a parley. Awad approached within speaking distance and opened a conversation with them, whilst Musa and I remained at a distance with our guns in readiness in case of a surprise or treachery. After the usual salutations had passed between them and some questions had been asked and answered, the Bedouins declared that they had no intention of molesting us, and told us that we might proceed on our way. They then begged for a little bread, saying they were hungry, and, in order to encourage us to approach, one of them dismounted and handed his spear to Awad. I advanced and, after saluting them, directed Antonio to give them some bread. We then sat down and ate and smoked together. I asked them their names, and noted them in my pocket-book, doubting, however, whether they had given them to me correctly.

Antonio declared that these Bedouins were enemies of his tribe, and would certainly have cut his throat if they had recognised him. He was almost paralysed with fear, especially when he overheard them, as he pretended, proposing to my guards to join with them in plundering me—for, they maintained, Franks never travelled with less than fifty purses in their pockets, and if they could only get rid of me they would be rewarded by a rich booty. However, my Arabs, either from fear of the consequences or because they were too mindful of the duties of hospitality to betray me resisted this appeal to their cupidity. After a short delay I remounted my camel, and we proceeded on our journey. But Awad, who was not satisfied as to the intentions of the Bedouins, warned me to be on my guard, as they might dog our steps and endeavour to take us by surprise. He lingered behind for some time on the watch, and it was only when he was satisfied that they were not following us that he rejoined me.

When he came up to me he lifted both his hands to heaven, exclaiming Allah! Allah! and drawing one of his fingers across his throat, to give me to understand what my fate would probably have been had I fallen into their hands.

Leaving the sandy plain, we entered a range of lofty hills, or rather mountains. The country continued to have the same savage and desolate appearance. It would be difficult to imagine a more wild and inhospitable region than that to the south of the Dead Sea. It is called the Wady

Ghor. Water is rarely to be found in it. The soil is barren and stony, and without vegetation. In summer the heat, radiated by the parched and burning ground, and the hot sultry atmosphere are almost fatal to life. No human habitation is to be seen, and no living creature moves during the day on the face of the earth.

In a sheltered valley we came unexpectedly upon a running stream, with a few bushes and stunted trees upon its banks—a sight that rejoiced man and beast, as we had now been for three days without water, except that which, on account of its brackish and noisome flavour, I could scarcely drink, and which rather excited than allayed thirst. The Arabs called the spot Fédan.

Some masses of rock, detached from the overhanging cliffs and blocking up the narrow valley, had formed, lying one above the other, a natural doorway about three feet high, through which my guards crept. This appeared to be a kind of religious obligation. The spot was evidently, for some reason or another, considered holy by the Arabs, as there was a heap of stones hard by to which every one who passed added, and pilgrims stuck in the crevices bits of rag torn from their garments—practices common in other parts of the East. Awad could only explain that a saint or dervish had once inhabited the place; but who this saint was and whence he came he did not pretend to know. He informed me that those who made a pilgrimage to the spot and invoked the holy man's aid when creeping through the opening were cured of every kind of disease. He even went so far as to declare, after he had performed this simple feat, that he suddenly felt himself greatly relieved from some rheumatic pains in the leg from which he had been long suffering.

After we had refreshed ourselves at this cool and grateful stream, we recrossed the hills, and found ourselves again in the sandy plain. My guards begged me to cook my dinner before dark, as a fire at night might betray our camping place. It appeared to me that the smoke might have the same result in the daylight. However, I stopped as they requested, and boiled some rice — my only food. By three o'clock we were again on our way. Awad and Musa sought, by deviating from the regular track and by keeping among the low hills, to conceal our movements from the Bedouins whom we had met in the morning, and who, they were persuaded, had not abandoned their intention of falling upon us, and had probably been joined by other horsemen. They would not permit me to pitch my tent, as it might be seen from afar, and although the night was bitterly cold I was compelled to lie, wrapped in my cloak, in the open air.

I had not much sleep, as we had to keep watch, and at midnight I had to remount my camel, as the guards, although we had had no alarm, would not be induced to remain any longer in a place which they considered specially dangerous An hour before sunrise we found ourselves again on the beaten track leading to Petra. I felt so much exhausted from want of sleep and the fatiguing motion of the camel, that after we had entered a sheltered valley I dismounted, and was soon in a deep slumber. I was much in need of it, as I had scarcely slept since leaving Hebron. Refreshed by this rest, and after eating a little bread, I commenced the ascent of a steep, rocky mountain, called, as far as I could learn, Gebel Memella. In the distance to the south could be distinguished the lofty summit of Mount Hor—the Jebel Harun, or Mount Aaron, of the Arabs—and to the north three remarkable granite peaks rose from the serrated range of mountains into which I had entered. One of them the Arabs called Abu-Sekakeen—the father of knives—from its sharp and jagged outline; another Gebel Nobak. The centre mountain peak was the one towards which I was ascending— the Gebel Memella.

After toiling up a very steep and stony track for about two hours, the camels, unaccustomed to such mountain ascents, appeared to be much fatigued. Leaving them to rest for awhile under the care of Antonio, I ascended on foot, with Awad and Musa, a high peak in the neighbourhood. The day being cloudless, I anticipated a fine prospect from it, and was not disappointed. The scene was wonderful, and magnificent from its savage desolation. Range after range of barren, naked hills of the most varied and fantastic shapes, like the waves of a sea which had been suddenly arrested when breaking and curling, stretched before me. Beneath me lay the inhospitable valley of the Ghor. In the extreme distance, to the north, could just be distinguished the Dead Sea. I took some observations with boiling water and my thermometer to obtain the approximate height of the peak, made some coffee, and descended to Antonio and the camels.

We continued, during the afternoon, the ascent of the remarkable and picturesque range of mountains we had entered in the morning. They were of sandstone, in strata of various colours, and, being very fragile and easily decomposed, had been worn into the most fantastic shapes, such as domes, pinnacles, and pyramids, which looked as if they had been the work of human hands. In some places these strata had the appearance of having been at one time in a liquid state, and to have flowed one over the other, and then to have hardened, like honey on the lips of a jar. This sandstone absorbs water so rapidly that, although it had rained heavily during the night before, we could only find one dirty, brackish pool from which to fill our water-skins. We now entered a long narrow gorge, formed on either side by lofty cliffs, broken into every variety of form. Through it ran the bed of a torrent, then dry and filled with trees and shrubs. I was desirous of pitching my tent as I saw many excavations in the rocks which I wished to examine; but my guards declared the place to be specially dangerous from robbers, and hurried me through it as fast as the camels could go. I had only time to make a hasty sketch of a sepulchral monument carved on a scarped rock about twenty-five feet above the level of the torrent bed. It consisted of two pyramids in high relief resting upon one base, upon which was a tablet with an inscription almost entirely effaced. Outside this tablet I could perceive a few Greek letters which had formed part of a second inscription.

There were probably other sculptured monuments in this gorge, but I could not stop to search for them. I observed in various places flights of steps cut in the rocks, once probably leading to tombs or habitations of which no remains, as far as I could ascertain, existed. When we emerged from this ravine we came upon a mountain declivity with a little vegetation. I pitched my tent for the night under a huge projecting rock by which we were completely concealed. As I sat by it in the calm evening large red-legged partridges[2] swarmed around me, loudly cackling and crowing. They offered tempting materials for an excellent supper, after the privations of the previous days, when my only food had been boiled rice and cakes of unleavened bread baked in the ashes; but my Arab guards implored me not to use my gun, as they were still haunted by the fear of robbers, and its report would disclose our hiding-place.

On the following morning we entered the Wady Musa, or Valley of Moses. I knew that I was approaching Petra by the innumerable monuments, chambers, flights of steps, and reservoirs, excavated in the precipitous rocks on either side of us. In an hour and a half I found myself amidst the ruins of the ancient city. Everywhere around me were remains of ancient buildings of all descriptions, whilst in the high rocks which formed the boundaries of the valley were innumerable excavated dwellings and tombs.

[2] The Red, or Greek, partridge—*Caccabis saxatilis.*

As I had intended to visit the ruins leisurely, I did not stop to examine them, but, passing through them on my camel, ascended to a spacious rock-cut tomb, in front of which I could perceive a small platform apparently covered with grass. There I made up my mind to pitch my tent.

I dismounted and spread my carpet. I had scarcely done so when a swarm of half-clad Arabs, with dishevelled locks and savage looks, issued from the excavated tombs and chambers and gathered round me. I asked for some bread and milk, which were brought to me, and Antonio prepared my breakfast, the Arabs watching all our movements. Their appearance was far from reassuring, and my guides were evidently anxious as to their intentions. They were known to be treacherous and bloodthirsty, and a traveller had rarely, if ever, ventured among them without the protection of some powerful chief or without a sufficient guard.

They remained standing round me in silence, until they perceived that I was about to rise from my carpet with the object of visiting the ruins in the valley. Then one of them advanced and demanded of me in the name of the tribe a considerable sum of money, which, he said, was due to it from all travellers who entered its territory. I refused to submit to the exaction, alleging that I was under the protection of Sheikh Abu-Dhaouk. I was ready, I added, to pay for any provisions that might be furnished to me, or for any service of which I might be in need.

This answer gave rise to loud outcries on the part of the assembled Arabs. They began by abusing my two guides, whom they accused of having conducted me to Wady Musa without having first obtained the permission of their sheikh. A violent altercation ensued, which nearly led to bloodshed, as swords were drawn on both sides. An attempt was made to seize my effects, and I was told that I should not be allowed to leave the place until I had paid the sum demanded of me. As I still absolutely refused to do so, one, more bold and insolent than the rest, advanced towards me with his drawn sword, which he flourished in my face. I raised my gun, determined to sell my life dearly if there was an intention to murder me. Another Arab suddenly possessed himself of Musa's gun, which he had imprudently laid on the ground whilst unloading the camels.

I directed Antonio to inform the crowd, which was now increasing in numbers, as men and women issued from the rock-cut tombs like rabbits from a warren, that I was under the protection of Sheikh Abu-Dhaouk, who had made himself personally responsible to the Governor of Hebron, and consequently to Ibrahim Pasha,[3] for my safety. If any violence were offered to me he would lose his head, unless his tribe took full vengeance upon those who had committed it, and the Egyptian Government would not be satisfied until they were exterminated.

The tribe inhabiting the Wady Musa had not long before been at war with that of Sheikh Abu-Dhaouk, who had inflicted considerable losses upon it, killing some fifty of its best warriors and carrying off a large number of its sheep and camels. It had reason, therefore, to fear that an outrage upon a traveller who was under his protection might lead to serious consequences. A consultation took place among those who appeared to have some authority over the crowd, which ended by my being informed that if I agreed to pay about half the sum at first demanded I should be allowed to remain as long as I liked in the valley, and to visit the ruins without molestation. What they asked, they declared, was far less than had been paid by other travellers, and it was only out of consideration for a guest and friend of Abu-Dhaouk that they would be satisfied with so small a sum.

[3] Ibrahim Pasha, the son of Mehemet Ali Pasha, was then in Syria at the head of the Egyptian army, and governed the country with a rule of iron.

I still refused. In the first place, I thought it right to resist this attempt to impose blackmail upon a traveller; and, in the second, had I been even disposed to accede to it, I had not enough money with me to give what was asked. I therefore directed Musa and Awad to reload the camels and to prepare to accompany me. Seeing that I was determined to carry out my intention of visiting the ruins without their permission, the Arabs formed a circle round me, threatening to prevent me from doing so by force, gesticulating and screeching at the top of their voices. With their ferocious countenances, their flashing eyes and white teeth set in faces blackened by sun and dirt, and their naked limbs exposed by their short shirts and tattered Arab cloaks, they had the appearance of desperate cut-throats ready for any deed of violence.

In this juncture, and when an affray which might have led to fatal results seemed imminent, the Sheikh of Wady Musa, who had been absent from the valley, made his appearance. Having somewhat calmed his excited tribesmen and obtained silence, he inquired into the cause of the disturbance. Having been told it, he announced that he had a right as chief of the tribe in whose territory the ruins were situated to the sum originally demanded, and that unless I paid it he would not permit me to visit them.

He was a truculent and insolent-looking fellow, tall, and with a very savage countenance; rather better dressed than his followers, and armed with a long gun and pistols, whilst they only carried swords and spears.

I repeated my resolution not to submit to this imposition, and warned him, as I had done his followers, that if any injury befell me he would be held personally responsible by Ibrahim Pasha, who had given ample proof that he could punish those who defied his authority. Abu-Dhaouk, moreover, I said, was a hostage for my safety. With these words I rose from my carpet and, directing Awad and Musa to follow me with the camels, which they were loading, prepared to begin my examination of the ruins.

The sheikh, seeing that I was not to be intimidated, and fearing the consequences should any violence be offered to me or to my guides which might lead to a blood feud between his tribe and that of Abu-Dhaouk, ordered his men to stand back, and I went on my way without further interference. As I descended into the valley he called out to me by way of benediction, 'As a dog you came, as a dog you go away.' I gave him the usual Arab salutation in return, and threw him a piece of money in payment for the bread and milk which had been brought to me on my arrival. This return for hospitality would have been resented as an insult by a true Bedouin, but he picked up the silver coin, and as I left the little platform in front of the tomb I saw him crouching down on his hams surrounded by his Arabs, evidently discussing the manner in which I ought to be dealt with.

Awad and Musa were a good deal alarmed at my reception, and feared that the sheikh and his followers would find some means of avenging themselves upon me for having defied them. They urged me, therefore, to leave the valley as soon as possible. But I was convinced that, notwithstanding the chief's threats, he would not venture to rob or injure me. The name of Ibrahim Pasha was at that time feared throughout Syria, and the sheikh could not but be well persuaded that Abu-Dhaouk, to save his own head, would execute summary vengeance upon those who had plundered or murdered a traveller under his protection. I was determined, as I had come so far to visit the ruins of Petra and its principal monuments, to examine them leisurely, and I spent the whole day in doing so. I was not molested, but I observed Arabs watching all my movements.

I had sufficient time to visit the remarkable ruins which Laborde and other travellers had described—the great amphitheatre carved out of the rock, the various temples and public edifices, and many of the tombs sculptured in the precipitous cliffs forming the sides of the valley. These tombs, some of which were elaborately ornamented with pediments, friezes, and columns, were mostly used as habitations by Arab families, and their spacious chambers were filled with smoke and dirt.

The scenery of Petra made a deep impression upon me, from its extreme desolation and its savage character. The rocks of friable limestone, worn by the weather into forms of endless variety, some of which could scarcely be distinguished from the remains of ancient buildings; the solitary columns rising here and there amidst the shapeless heaps of masonry; the gigantic flights of steps, cut in the rocks, leading to the tombs; the absence of all vegetation to relieve the solemn monotony of the brown barren soil; the mountains rising abruptly on all sides; the silence and solitude, scarcely disturbed by the wild Arab lurking among the fragments of pediments, fallen cornices and architraves which encumber the narrow valley, render the ruins of Petra unlike those of any other ancient city in the world. But I felt somewhat disappointed with the ruins themselves, of which I had read such glowing descriptions.

I thought the architecture debased and wanting both in elegance and grandeur. It is of a bad period and of a corrupt style.

The most striking feature at Petra is the immense number of excavations in the mountain-sides. It is astonishing that a people should, with infinite labour, have carved the living rock into temples, theatres, public and private buildings, and tombs, and have thus constructed a city on the borders of the desert, in a waterless, inhospitable region, destitute of all that is necessary for the sustenance of man— a fit dwelling-place for the wild and savage robber tribes that now seek shelter in its remains. However, if a city were to be constructed on such a site, it was far easier and much less costly to carve the soft and fragile rock into temples and habitations than to bring stone fit for solid and durable masonry from a distance. The remains of a bridge prove that the torrent whose bed occupied the centre of the valley was not always dry. Water was collected in reservoirs excavated in the rocks; but it must have been brackish and unpleasant to the taste, judging from that which we obtained from the pools, and the Arabs, I was assured, used no other.

I made notes and a few rough sketches of the principal remains, and towards evening, yielding to the urgent entreaties of Awad and Musa, who declared that it would not be safe for me to pass the night among the ruins—for if we were not attacked we should certainly be robbed—I left them by the valley through which we had arrived. We encamped some time after nightfall in a narrow wady. But we had little sleep, as we had to keep watch, fearing that the Arabs of Wady Musa might have followed us with evil intent.

CHAPTER III.

Leave Petra—Danger from Arabs—The Wady Ghor—The Seydi'in Arabs—The Dead Sea—The Mountains of Moab—Plundered by Arabs—Arrival at Kerak—The son of the Mujelli—Recovery of my property—The Christians of Kerak—Its ruins—My Christian host—Sheikh Suleiman Ibn Fais—Departure for the Desert—Encampments of Christian Arabs—The Country of Moab—The Beni-Hamideh—Remains of ancient towns—Meshita—An Arab banquet—The tents of Suleiman-Ibn-Fais—Reach Ammon.

WE were on our way again long before daylight, and, crossing the same desolate and arid sandhills and ravines which we had passed on our way to Petra, stopped for the night on the stream of Wady Fédan. During the whole day's journey Musa and Awad had been in sore fear of robbers, and never ceased praying for our safe arrival at Kerak, vowing to sacrifice a sheep on their return to their tents if no accident befell us. These two men proved unusually good specimens of the Arabs who, encamping on the borders of Palestine, have been corrupted by contact with the Turks and by injudicious travellers. They served me zealously and faithfully, and did their best, in the face of many difficulties and no little danger, to see me safely through my venturesome expedition. They were handsome, well-made men, with open, intelligent countenances, hardy and able to bear great fatigue, and, as far as I could judge by their conduct, by no means deficient in courage. I felt that in the somewhat perilous position in which I was placed I could trust them.

Although the fear of attack from the Arabs of Petra had somewhat decreased, my guides would not permit me to travel after dark in the sandy plain upon which we were about to enter, on account of the venomous serpents which, they vowed, came out from their holes at night, and were much to be dreaded by those who went on foot. Next morning they pointed out to me what they assured me were the traces of these reptiles in the sand.

We followed the valley of El Ghor during the day, in the direction of the Dead Sea. The chain of the mountains of Moab, their summits covered with snow, bounded the horizon to our right. Ranges of sandstone, limestone, and granite, each with its peculiar features, ran parallel to each other—the fragile sandstone, yellow and barren, worn by the weather and by winter torrents into fantastic shapes; the limestone, with undulating forms and some verdure on the slopes; the granite, with lofty peaks and serrated outlines.

As we approached, on the following morning, the foot of the high land of Moab, which rises to the east of the Dead Sea, we came upon rivulets of fresh water and grass. In the distance were flocks of sheep and camels grazing. After crossing three streams, called respectively by the Arabs Wady T'lah—near which were the ruins of an extensive building—the Wady Khanaisir, or the valley of wild boars, and the Wady Féfé, we arrived at an Arab encampment. The black tents, pitched near a clear stream and in high brushwood, which concealed them entirely from view until we were almost in the midst of them, belonged to the tribe of Seydi'in, under Sheikh Mahmoud Abu-Rueri. These Arabs were known to my guides, who suggested that I should spend the remainder of the day with them, promising me a hospitable reception. I accordingly pitched my tent, and, having purchased a sheep for twelve piastres, ordered half of it to be at once boiled. Neither my companions nor myself had seen meat since we had left Sheikh Abu-Dhaouk's tent. I had eaten of the rice which I carried with me; but they had lived entirely upon cakes of unleavened bread which they had baked in the hot ashes. We were consequently happy at the prospect of a meal which would satisfy our hunger. But it was from the want of water fit to drink that I had suffered most, and I drank with exquisite delight from the bright sparkling stream which descended from the overhanging mountains of Moab.

My guards were not sorry to give their camels food and rest. The poor animals had been living for several days upon nothing but what they could pick up in the sandy plains and barren hills amongst which we had been wandering—an occasional tuft of camel-thorn and a low bush bearing a yellowish berry. They were now supplied with some chopped straw. Awad said that, although they could go for two or even three days without water, they ought to drink once in every twenty-four hours.

As it rained in the evening, I moved into one of the Arab tents—my own not keeping out the wet. A large wooden bowl filled with camel's milk, flour, melted rancid butter and pieces of unleavened bread, was brought to me for supper. The men made balls of this mess with their dirty hands, and then presented them to me to eat—a mode of showing attention to a guest which would not be altogether agreeable to a person choice as to his food. The tent was filled with Arabs who came to look at the stranger and to gossip. An open space in the midst of the encampment was reserved for the camels, which were driven into it at sunset, and made to kneel down, grunting and growling the while. The tribe had above a thousand of these animals, but few sheep or goats. The value of a camel, I was told, was from six to eight hundred piastres (from 6*l.* to 8*l.*).

Late at night a Bedouin arrived with the news that two tribes in the neighbourhood had fought in the morning, with the loss of a few men on either side. Amongst the killed was a relation of my guide Awad, who expressed his grief at his death in loud lamentations, interrupted by curses on the man who had slain him, and vows of vengeance. The affray had arisen in consequence of a blood-feud.

The black camel-hair tents of these Dead Sea Arabs were similar to those with which I afterwards became so well acquainted among the Bedouins of Mesopotamia. They were spacious, supported by a number of poles, and divided into two parts— one for the men and for guests, the other for the women and children and for domestic purposes, such as cooking and baking. Even in the encampments of these debased Arabs a tent was usually set apart for guests. In it during the cold winter nights a blazing fire was constantly kept up, round which the idlers gathered to look at the traveller, and to learn the news of the day.

The Arabs I had hitherto seen were fine and well-proportioned men. But their sparkling eyes, their white teeth, which they constantly showed, their long tails of plaited, well-greased black hair, their swarthy complexions and freely exposed limbs, gave them a savage and forbidding appearance. The women, as far as I could judge of them, were less handsome than the men. They took no great care to conceal their faces, only the lower part of which they slightly covered by a piece of dark linen of a triangular shape, which was fastened by a cord or by a metal chain to a headband or turban. Round their arms and ankles they wore many bangles of silver, and of the blue opaque glass made at Hebron. The children of both sexes were generally without clothing of any kind. The tents were dirty and abounding in vermin.

Ibrahim Pasha had succeeded in depriving most of the Arab tribes living near the Dead Sea of their firearms; but had failed to establish his authority permanently over them, although they had consented to pay a small annual tribute to the Egyptian Government. They were consequently, with few exceptions, only armed with swords, knives, and spears. Some had not even these weapons, but only thick wooden clubs with a big knob at the end.

I had no reason to complain of our hosts, who treated us hospitably, and did not clamour for money as others had done. We left them early in the morning, and, descending to the shore of the Dead Sea, continued along it by a very rough and stony track, leading over the detritus washed down by the water torrents from the mountains, which rise almost abruptly from the water's edge. We crossed several of these torrents, which were almost dry. In some the water was sweet and agreeable to the taste, in others brackish and scarcely drinkable. I obtained from the Arabs their names, and noted them in my journal.[1]

Shortly after crossing the Nahr Assal we fell in with a tribe changing their encampment and seeking for fresh pastures. They were accompanied by large flocks of sheep and goats and herds of camels. Their sheikh was a handsome young man, named Ibn-Rashid. He had on his saddle before him his little son, whom he gave over to one of his attendants soon after meeting me, saying that he meant to accompany me until he had seen me safely through his people. He was

---

[1] I find the following—Kasr el Ghor Dhafyeh, with sweet water; El Gherachi, also sweet water; El Murah, bitter or brackish water, as the name denotes; Nahr Assal, or the honey stream, &c.

courteous and obliging, and did not ask for a present, which surprised me. He was surrounded by a number of horsemen carrying spears tufted with ostrich-feathers, and by armed men on foot. Amongst his attendants was a huge and ferocious-looking negro, wearing a long robe of red silk and a white turban, which added to the hideousness of his countenance.

After I had parted with the sheikh and his attendants I dismounted and bathed in the Dead Sea. I found the waters as buoyant as previous travellers had described them to be, but their extreme saltness caused an unpleasant irritation of the skin. When I was about to resume my journey Musa begged me to permit him to leave me. He said that there was a blood-feud between his family and the Arabs in the neighbourhood of Kerak. They would certainly recognise him, and he would be killed. He proposed to wait in the tents we had left in the morning until Awad, who had no quarrel with the Kerak Arabs, returned with the two camels. I parted with him with regret, as I believed him to be trustworthy, and thought I could rely upon his courage and presence of mind in the event of danger. I was well satisfied with the services of Awad, but he appeared to me to have less tact and judgment than his companion.

Awad, who, notwithstanding his miraculous cure at Fédan, was still suffering from rheumatism in one of his legs, got up behind Antonio on his camel. He suddenly slipped off the animal's back and disappeared amongst the reeds of a marsh between our track and the Dead Sea. Returning after some time, he explained his disappearance by declaring that he had seen some one lurking among the rushes who had fled on being discovered. I was disposed to treat his suspicions lightly, but as we continued our journey I perceived a man on the rocky declivity above us, who was evidently watching our movements with no good intentions. Awad was so persuaded that we were about to be attacked and robbed that, after we had pitched my tent in a narrow gully, he remained on the alert all night.

Nothing, however, occurred to disturb my rest. An hour before sunrise I was again on my camel. We now began to ascend rapidly, through a valley called 'Wady Dea,' in the direction of Kerak. After reaching a considerable elevation, which commanded a beautiful view of the Dead Sea and the opposite heights towards Jerusalem, the camels being much fatigued by the steep and rocky precipitous ascent, I directed them to be unloaded and some breakfast to be prepared, whilst I walked to a higher point from which I could better enjoy the prospect.

We had been joined by an Arab on foot, whom Awad recognised as one Mahmoud, the sheikh of a small tribe encamping in the neighbourhood of Kerak. He was accompanied by several men, who had remained at a distance from us. On my return to breakfast I perceived this man struggling with Awad for my saddle-bags. I seized him and asked for an explanation of his conduct. His answer, as translated to me by Antonio, who was trembling with fear, was to the effect that he had many followers, and that unless I at once gave him a considerable sum of money as bakshish for passing through his tribe he would cause me to be robbed and murdered. Having thus delivered himself, he relinquished his hold upon my saddle-bags and was going away, when, as Awad sought to detain him, the struggle between them was renewed. Thinking that he might carry out his threat of attacking us, I determined to keep him as a hostage. I pointed my gun at him and threatened to shoot him if he attempted to leave us. In abject fear, as my gun was levelled at his head and not far from it, he begged for mercy. I directed Antonio to disarm him, and in his fright he gave up his pistols, knife, and club without attempting to offer any resistance.

Having disarmed the sheikh, I sat down on the

ground to eat my scanty breakfast of boiled rice, keeping my eye upon him the while and my gun ready across my knees. I then invited him to eat bread with me. He at first refused, but ended by dipping his hand into the dish. The camels were reloaded, and I directed him to lead the way to Kerak, repeating my threat to shoot him if any attempt were made to attack and plunder me on the way. That he might be persuaded that I would do so, I walked close behind him.

In about an hour we reached a part of the ravine in which there was an Arab encampment below us and another above. Mahmoud wished to take me to the upper tents, saying that he wanted water to drink, and that as they belonged to his people I should be his guest and he would kill a sheep for my entertainment. As I suspected that he wished to get me into his hands, I ordered him to continue on the direct track to Kerak.

I was walking behind him, about two hundred yards in front of the camels, when two Arabs, armed with spears, came running down the slope. I directed Mahmoud to order them back. They, however, approached Awad, and whilst one of them gave him the usual salutation the other snatched away my cloak, which I had left on my pack-saddle, and ran off with it as fast as his legs could carry him. Mahmoud offered to pursue the thief—an offer which I peremptorily declined to accept.

This first theft was the signal for a general attack upon my property. Arabs appeared, as if by magic, from above and below. I dragged Mahmoud, whom I had seized by the arm, towards the camels. Thinking that I intended to shoot him, he entreated the robbers, who had almost surrounded me, to draw back. Seeing the danger to which their sheikh was exposed, they hesitated to fall upon me. A wild-looking fellow, whose name I afterwards learnt was Beshire, however, approached me, menacing me with a spear which he raised above his head as if to throw at me. Awad seized him by the arm. A struggle ensued between them, in which his 'keffiyeh '— the kerchief which the Bedouins wear on their heads—fell off. With his many plaited tails of black hair, half concealing his face, and with his expression of mingled stupidity and ferocity, and his loud, guttural cries, he seemed a very devil. Releasing himself from Awad, he sprang upon an overhanging rock and again raised his spear, as if about to hurl it at me, swearing at the same time that he would have my blood.

Dragging the sheikh after me, and with my gun still at his head, I gained a small open space out of Beshire's reach. The other Arabs, fearing for their chief, who was loudly calling out for mercy, still hesitated to fall upon my camels and my little baggage. At this juncture the sheikh of the tents which were in the valley below—the encampment above belonged, it now appeared, to Mahmoud—arrived on the scene. He approached me unarmed, and appeared to be desirous to put a stop to the attack upon me. Awad, who was known to him, explained that I was travelling under the protection of Abu-Dhaouk, who was responsible to Ibrahim Pasha for my safety, and that I had letters for the Mujelli, or governor, of Kerak, who would also be answerable to the Egyptian Government should any injury befall me.

This explanation appeared to have a good effect. The Arabs sheathed their swords, which they had drawn when the *mêlée* began, and I expected to be allowed to continue my journey without further molestation, when the sheikh who had interfered in my behalf asked me to deliver up my gun. This I refused to do, and we were discussing the matter, which promised to be amicably settled, after he had read the letter to the Mujelli, when the ferocious Beshire, who had been watching his opportunity, suddenly threw himself upon Awad and seized his gun. A struggle ensued, the issue of which I watched with some anxiety, determined to shoot Beshire should he succeed in possessing himself of the weapon. In the excitement of the moment, I had relaxed my hold upon Sheikh Mahmoud, who endeavoured to make off. But running after him, I again seized him by the arm, and, holding my gun at his head, dragged him to some distance.

The struggle between Beshire and Awad led to a renewal of the affray. The Arabs began throwing stones, and Beshire, who had not succeeded in wresting the gun from Awad, threw his spear at me. Fortunately it glanced by me. My assailants again drew their swords, and one or two fired their pistols at me; but they were too far away to reach me. I exerted myself with all my might to drag Sheikh Mahmoud towards Kerak, menacing him with death when he attempted to stop. The camels had become restive with the noise and confusion, and had turned back, so that there was, by this time, some distance between me and them. Some Arabs followed them, and began to plunder my effects. The others, probably fearing that they would not have their share of the booty, instead of pursuing me, joined their companions, and I could see them, as I hurried on with the sheikh, dividing the contents of my saddle-bags and my other property.

Antonio had taken to his heels, and was calling

to me to follow him, but I felt that my safety depended upon keeping my hold on Sheikh Mahmoud, whom I hurried onwards as fast as I was able.

The Arabs, having taken possession of my effects, over which they were probably quarrelling, did not seem disposed to renew the attack upon me. After a long and toilsome walk I came in sight of an encampment not far distant from the road. Antonio, who had left me in his fright, expecting that the Arabs would pursue us, returned to meet me, and informed me that the tents belonged to some dependents of the Mujelli of Kerak, and that 1 should be in safety if I reached them. I soon afterwards perceived Awad driving the two camels before him. When he came up to me and confirmed what Antonio had said, I released Sheikh Mahmoud, who lost no time in making the best of his way back to his friends.

The sheikh of the encampment, to whom I showed my letter to the Mujelli, received me civilly, invited me into his tent, and ordered coffee to be prepared for me. I was soon surrounded by a group of curious Arabs, to whom Awad related, in all its details, my morning's adventure. They congratulated me upon my escape, the Arabs who had plundered me being, they said, the most notorious robbers and cut-throats to the east of the Dead Sea. I owed it entirely, they were convinced, to the fact that Sheikh Mahmoud had been in my power and was afraid for his life, and they examined with the greatest interest my double-barrelled gun, which had inspired him with so much fear.

The sheikh offered to conduct me in person to Kerak. Leaving Awad with the camels to take the beaten track, we ascended on foot the precipitous side of a mountain, on the summit of which we could distinguish the ancient walls of the castle frowning over the valley. We entered it by a long, narrow, vaulted passage cut through the rock, and I found myself in the midst of a mass of ruins through which we had to make our way to reach the house of the Mujelli. He was absent, but we were received by his son, a handsome youth, with ringlets of black hair falling from beneath his keffiyeh. I placed in his hands the letter from Colonel Yusuf Effendi to his father, and, relating what had befallen me, asked for his assistance in recovering my stolen property. The room in which he received me was small and dirty, and crowded with Arabs. The letter was read and discussed, and great indignation expressed at the conduct of the people who had threatened my life and plundered me, especially when it became known that Sheikh Mahmoud had eaten bread

with me in the morning. It was decided that
Ahmed, the young chief, should himself proceed at
once to the tents of the robbers with some horsemen
to recover my effects, even by force if necessary.
His mare was soon saddled and brought to him,
with his gun, sword, and pistols. A number of his
attendants were ready to accompany him, and he
requested me to go with him to identify my property. A mule, with a roomy pack-saddle, was
procured for me, and I mounted it with Antonio
behind me. I carried my gun, thinking that it
might prove useful should the Arabs offer any
resistance.

After we had issued from the castle, Ahmed
dismounted to recite his afternoon prayers, and
we then descended rapidly to the encampment of
Sheikh Mahmoud. Giving our beasts to some men
who came out to meet us, we entered the tent of
the sheikh who was there to receive us. It was
soon filled with Arabs, amongst whom I recognised
some of those who had attacked me in the morning.
Coffee was made and handed round, and we sat
smoking our chibuks for about half an hour
without any words being exchanged except the
usual salutations. As far as I could judge from
the countenances and appearance of the men, I
argued ill of the attempt of Ahmed to recover my
property, for I had never set eyes upon a more

ferocious, forbidding set of ruffians than those who were glaring at me.

At length Ahmed, having drank his coffee, smoked his pipe, and duly rested himself, addressed the assembled Arabs in a set speech, which, if Antonio's translation could be understood and trusted, was to the following effect. This Frank, said he, had been the guest of Abu-Dhaouk, and was under the protection of that sheikh, who was the friend and ally of the Arab tribes of Kerak. Men of the Jehalin [2] had been sent with him so that he might go his way in security. He had, moreover, letters from the officers of the great Ibrahim Pasha, recommending him to the Mujelli of Kerak, of whom he was consequently to be also considered the friend and guest. He had passed in safety amongst the robbers of Wady Musa and El Ghor, but when within two hours of the place where he expected to be welcomed and hospitably entertained he was robbed, and his life threatened, by those who ought to have honoured him as the Mujelli's guest, and to have given him help. And, what made matters still worse, Sheikh Mahmoud, the chief of these malefactors, had partaken of this stranger's bread, and it was the sacred duty of the Arab to protect at all cost him with whom he had eaten. He concluded his oration, which

[2] Abu-Dhaouk's tribe was so called.

was delivered with the accustomed Bedouin eloquence and gesticulation, by exclaiming, 'This man walks with God whilst you walk with the devil!' and calling upon them to restore my stolen property without delay.

When he had ended the Arabs turned upon Sheikh Mahmoud, crying out, 'Why did you not tell us that you had eaten bread with this Frank, instead of joining with us in robbing him?' He seemed somewhat ashamed of himself, and began by denying that he had either eaten my bread or been in any way concerned in the robbery; but, finding that the evidence against him on both points was conclusive, he pretended that I had prevented him from interfering in my favour by seizing him and threatening to shoot him. Had I released him, he declared, he would have explained to his people that I was under his protection, and I should not only not have been molested, but should have been welcomed as a guest in his tents.

After a stormy discussion it was decided that, under the circumstances, such part of my property as could be found in the encampment should be restored to me; but the remainder being in the hands of the Arabs from the tents lower down in the valley, who had joined in robbing me, I must seek it from them. Sheikh Mahmoud then insisted that it would be an insult to him were we to leave without partaking of his hospitality, and a sheep was slain for our entertainment. Men were then ordered to collect from the tents and to bring any articles that might be found in them belonging to me. In the meanwhile a young Arab was brought to me who was writhing from excruciating pains in his stomach. I was told that it was in consequence of having swallowed some liquid he had found in my saddle-bags, which had poisoned him. I was entreated to cure him by giving him an antidote. Observing that he was no doubt punished by Allah for having robbed me, I pointed out that unless I knew what he had drank, and until I had my medicines, which were contained in a little case which had no doubt fallen into his hands, I could do nothing for him, and that he would probably die. His mother, who had accompanied him, then disappeared, and returned immediately after with my small medicine-chest and a bottle containing a mixture with creosote, which had been given to me at Jerusalem to assuage the pain I was suffering from the toothache. He had taken a gulp from this bottle, believing that it contained some kind of Frank brandy. I administered an emetic, which soon had the desired effect; but I inwardly rejoiced that the fellow had been well punished.

My things came in one by one. Those that were

not returned were of little value My money, watch, and compass, which were of the most importance to me, were on my person, and had consequently been saved. Antonio had snatched my small carpet from the camel when we were first attacked, and had carried it off on his shoulders. My little tent had not been taken by the Arabs, as it was probably of no use to them. The first object that appeared after the medicines was Antonio's sack, which contained all his worldly goods, consisting of a few old clothes, from which, however, one or two articles had been abstracted. Next came one half of my saddle-bags with my maps and notebook, and other things which I was not sorry to recover. Ahmed took them out one by one, examined them curiously, and then handed them round for general inspection. The other half of the saddle-bags, containing some clothes and my Levinge bed, had not been found, but I was promised it for the following morning.

It was now late, and large wooden platters were brought into the tent, with the boiled flesh of the sheep which had been killed in our honour cut into bits and placed, hot and smoking, upon a mess of flour and bread soaked in the gravy. The hungry Arabs gathered round them, and the meat disappeared in a very short time. Melted butter, very rancid, was then poured over the kind of paste which remained, and which the Arabs kneaded with their hands into balls. These they handed to their guests or swallowed.

I had to pass the night in the crowded tent in a very uncomfortable position. My cloak had not yet been restored to me, and the air was very cold and keen, notwithstanding a blazing fire which the Arabs kept up. I could not sleep much, as I could but reflect upon the events of the day and the perils I had run. From what I had learnt I was persuaded that if it had not been for the fear of my gun, and from having had Sheikh Mahmoud in my power, I should have been murdered by these lawless robbers. It was fortunate that I had not been under the necessity of using it, as if blood had been once spilt my life would unquestionably have been taken. But I was now entirely in the hands of the people of Kerak and of the Mujelli, who bore a very evil reputation, and were not to be trusted. It seemed to me very doubtful whether I should be able to reach the ruins to the east of Jordan without running very serious risk, whilst, on the other hand, an attempt to return to Jerusalem would be attended, after what had occurred, with considerable danger. At length, exhausted, I fell into a doze, from which I was, however, soon roused by the Arabs, who were on foot long before daylight.

As the sheikh had promised, the other half of my saddle-bags was brought to me early in the morning, but emptied of the greater part of its contents. I was assured that nothing whatever remained belonging to me in the tents, and that anything which I missed would be found in the other encampment. We accordingly descended the mountain side to it, after having drunk the usual coffee. The sheikh who had offered to protect me when attacked by the Arabs on the previous day, and had endeavoured to persuade me to deliver up my gun, received us. Ahmed addressed him and his followers in a speech somewhat similar to the one he had made to Sheikh Mahmoud. But it had not the same effect. The Arabs who had assembled scowled at me, and seemed disposed to resent his interference. Amongst them I recognised Beshire, the stupid savage who had thrown his lance at me. 'Who is this man,' he exclaimed, 'who calls himself a Frank and the guest of Abu-Dhaouk? How do we know that he is not a Jew? By what law is the Arab compelled to restore that which he has once taken? Are we not entitled after what has happened to have his blood? And who is the Mujelli and his son that we are to obey them and to give up at their bidding that which of right belongs to us?' At these words an old man, who had been very active during the attack upon me the day before, drew his sword, and flourishing it over his head, declared that he would retain what he possessed of my property whether the Mujelli's son wished it or not. Other bystanders also then drew their swords, and all of them began gesticulating violently and screaming at the top of their voices like madmen.

During this scene Ahmed remained silent and apparently indifferent, quietly smoking his short chibuk. When the noise had somewhat subsided, he ordered one of his attendants to bring him his horse, and was preparing to leave without making any answer to the speeches in which his authority and that of his father had been defied. The sheikh of the encampment, who had continued seated by his side, and had hitherto taken no part in the tumult, now interposed. I had come to Kerak, he said, under the protection of the sheikh of the Jehalin, and as the guest of the Mujelli. Had this been properly explained to his tribe I should not have been molested. What had occurred had been the consequence of a misunderstanding, and my property ought therefore to be given back to me. The majority of the bystanders appeared to agree with him. After an animated discussion the old man who was the first to draw his sword, which he had continued to flourish at me in a menacing manner, was induced to sheath it, and Beshire was compelled to yield. He went sulkily to his tent, and returning soon after with my cloak threw it in an insolent manner to me, and, swearing by the Prophet that it was all he had belonging to me, retired, muttering curses upon me. Various articles of clothing appeared one by one and slowly; but some of them had already been turned to account by the robbers. My second shirt had been cut up into three to clothe some naked children, and an Arab had ingeniously converted my only spare trousers into a jacket.

It was midday before a mess of bread soaked in gravy and melted butter was brought to us for breakfast, and as it became evident that I should be unable to recover the few articles of little value which were still missing, I proposed to Ahmed to return to Kerak, well pleased that I had succeeded through his assistance in getting back the things which were of most importance to me, such as my medicines, maps, and books, and a few indispensable articles of clothing. Before leaving the tents I learnt that the three horsemen I had met in the desert before reaching Petra belonged to this tribe. They were returning from a plundering expedition, and would have followed and attacked me in the night had I not written down their names. This appeared to have alarmed them, as they fancied that I must have had some mysterious object in doing so.

As we rode back to Kerak Ahmed informed me that the Arabs who had plundered me belonged to the tribe of Aranat; that they were notorious robbers; that they refused to pay taxes or tribute; and were only kept in something like subjection by their fear of the superior force of the Mujelli, who had on more than one occasion inflicted severe punishment upon them for their misdeeds.

I was sent by Ahmed to lodge with a Christian named Ibrahim—an obliging fellow, who sought to do his best to entertain me, but who was miserably poor. His dwelling was a wretched hovel, formed out of part of a ruined house, dirty and swarming with vermin. The Mujelli's son presented me with the carcass of a small lean sheep, which my host boiled and of which he was glad to partake, as he had scarcely even a bit of dry bread to give me. There were no provisions to be obtained in the town, which was a mere heap of ruins. The few Christian families of the Greek faith who still lingered among them were reduced to an almost starving condition. Ibrahim pathetically described their sufferings to me. Neither their lives nor the little property they possessed were secure, and, whilst cruelly oppressed and taxed within the walls, they dared not venture outside of them for fear of being murdered or plundered by the lawless tribes which inhabited the surrounding country.

Kerak had submitted to Ibrahim Pasha soon after the Egyptian occupation of Palestine. A small garrison was placed in the fortress. The Arab inhabitants of the town rose suddenly one night and murdered, it was said, four hundred and sixty soldiers who were quartered upon them. About six hundred Egyptians still held the castle. After a short time, being in want of provisions, they surrendered, a formal undertaking having been given to them that they should be allowed to retire to Jerusalem. They were treacherously attacked in the night during their retreat, when encamped at the entrance of the Wady Kerak, and the greater number of them were slain.

The Christians, who had taken no part in the massacre of the Egyptian troops, but who feared the indiscriminate vengeance of Ibrahim Pasha upon the inhabitants of Kerak, fled from the place to Hebron, abandoning their property, which was appropriated by the Arabs.

Ibrahim Pasha sent a force to punish the insurgents, who were able to resist for some time owing to the great strength of the position. The fortress, into which the Arabs had retired with their chiefs, was at length retaken, but not without considerable loss to the troops, who revenged themselves by slaughtering a number of their prisoners. The then Mujelli was captured with his two sons, and sent to Jerusalem, where, with the eldest, he was beheaded in the corn market. His brother, the father of Ahmed, was appointed his successor, on condition that he and his tribe would acknowledge the authority of the Egyptian Government and pay a small annual tribute. The Egyptians retired after destroying the principal part of the walls and fortifications, and the surrounding vineyards and olive plantations. At the time of my visit there was no garrison in Kerak, nor any representative of the Egyptian Government. The Christians, who had been deprived of their arms, returned to the town, but were unable to recover the property which they had left behind them, and being nevertheless compelled to pay taxes, like the other inhabitants of the place, they were reduced to a state of the utmost wretchedness. They were then about three hundred in number, and possessed a small church served by a solitary priest. The whole population only amounted to between eight and nine hundred souls, so much had it suffered in its struggle with the Egyptians. Such was the recent history of Kerak, according to Ahmed and my Christian host.

I spent the greater part of the day after my arrival in visiting the ruins, accompanied by Ahmed. I was greatly struck by the commanding position of the castle, and by its massive walls, which had resisted the attempts of the Egyptians to destroy them. One tower constructed of solid masonry still remained uninjured. The rock upon which it was built had been scarped, and in parts the mountain side had been coated with slabs of stone, like the artificial mounds upon which stand the castles of Aleppo and Harem. The only access to it had been by the vaulted passage cut through the rock, by which I had entered the town. The position, which is not commanded from any of the surrounding heights, must have been one of great strength. Amongst the ruins I discovered the remains of an ancient Christian church, upon the walls of which were still to be seen some rude religious paintings.

I was anxious to proceed without delay on my journey. But Abu-Dhaouk's camels had returned to his tents with Awad, and I was unable to obtain others to take me to Rabbath Ammon and Jerash, the two places which it was my principal object to visit. Those who possessed any, or who had horses or mules, refused to hire them to me. The desert through which I should have to pass was exposed, they said, to constant inroads from the Bedouins, and without the protection of one of their sheikhs and an escort of horsemen, it would be impossible for me to travel in it. I should be plundered and perhaps murdered, and they would lose their camels. I soon discovered that there was an intrigue at the bottom of these difficulties. Ahmed, who had professed to be my friend and to consider it a sacred duty to recover the property of his father's guest, was not so disinterested in the matter as he appeared to be. If he had enabled me to obtain from the Arab robbers the articles of which they had robbed me, it was only with a view to securing the best part of them for himself.

He began by begging for money, which I refused to give him. He then suggested that I might make him a present of my gun, and on my telling him that I would not part with it, asked for my pistols instead. Being still unsuccessful he wanted my sword, then my carpet, and, lastly, my cloak. He had some good reason to give for each of these requests. He was going to be married, and wanted a carpet such as mine. He had to make a present to his brother, and my sword would exactly suit him. Finding that he could get nothing out of me, he tried to extort something from Antonio, asking in turn for his 'tarbush' or red cap, his jacket, and his sash. The boy, alarmed at finding himself at the mercy of the Kerak Arabs, whom he greatly dreaded, was about to give way; but I peremptorily ordered him not to part with any of his garments.

In the meanwhile Ahmed had secretly forbidden the inhabitants to furnish me with camels or horses. Hence the difficulty I experienced in obtaining them. I learnt these intrigues from my Christian host, who behaved loyally to me, and gave me all the help in his power. A whole day was spent in wrangling and quarrelling. High words passed between Ahmed and myself. I reproached him for conduct so unworthy of an Arab who had boasted of his respect for the sacred rights of a guest, and ended by threatening to return at once on foot to Hebron, and to lay a complaint before Colonel Yusuf Effendi, holding him personally responsible for anything that might befall me on the way.

Finding that I was determined not to yield to the imposition which he sought to practise upon me, and fearing that his conduct towards me might bring his father into trouble with the Egyptian authorities, he came to me in the night to inform me that he expected one Suleiman-Ibn-Fais, a sheikh of the Beni-Sakk'r Bedouins, to pass through Kerak on the following day, on his return to his tents, which were probably pitched among the ruins of Ammon or in the neighbourhood. He would place me under his protection, he said, and would obtain his consent to my accompanying him. I should have to find the means of reaching the encampment of the sheikh, who, when we had arrived there, would no doubt provide me with camels to perform the remainder of my journey. He offered to hire two mules for me, for which I should have to pay 120 piastres to the owner, with whom, I afterwards learnt, he was to divide the money. To avoid further altercation I consented to give this sum, and the mules were to be ready on the arrival of the Beni-Sakk'r chief.

Sheikh Suleiman-Ibn-Fais, who arrived as Ahmed had announced, was a tall, handsome man of very dignified appearance, with regular features, bright intelligent eyes, and a long bushy black beard, such as is rarely seen amongst Arabs. He wore the keffiyeh, from under which fell several long plaits of black hair. Under his striped Arab cloak was a robe of rich Damascus figured silk. In his girdle he carried a pair of silver-mounted pistols.

The sheikh stopped to eat bread at the house of Ahmed, who introduced me to him. He received me courteously and offered to take me to his encampment, which, as I had been informed, was near the ruins of Ammon. I should be his guest, and he would provide me with camels for the rest of my journey, which, under his protection, I should be able to perform with perfect safety. When the time came for our departure, instead of the two strong mules that I had been promised, I was offered one half-starved animal for myself and a donkey for Antonio. I remonstrated, protesting against this fresh act of dishonesty and duplicity on the part of Ahmed. But he vowed that he had failed in his endeavours to obtain other animals, as the inhabitants of Kerak had been so completely pillaged by the Egyptians that they no longer possessed a horse, camel, or mule amongst them. As Suleiman-Ibn-Fais was in a hurry to leave, I was under the necessity of yielding, fearing to lose the opportunity of accompanying him.

Amidst all the troubles and vexations I suffered from the knavery of the son of the Mujelli, it was not a little pleasing to me to experience from the poor Christian, Ibrahim, a disinterested kindness and hospitality which were quite unexpected. When I offered to pay for my entertainment he absolutely refused to receive any remuneration whatever. Nor would his wife take the money I offered her. It was even with much difficulty that I prevailed upon him to allow me to give his little son some small silver coins. When the flour and butter which Ahmed had promised to procure for me as provisions for my journey were not forthcoming, the worthy pair insisted upon my accepting a small bag of the former out of their scanty store. They appeared to sympathise sincerely with my somewhat forlorn condition, and warned me of the danger that I was running by putting myself in the power of people whom they believed capable of every perfidy and crime. When I left them they devoutly offered up prayers for my safety, and earnestly entreated me to be on my guard during my journey.

Before leaving Kerak I voluntarily gave Ahmed the sword which I wore, and my tent, as I considered that he was entitled to some acknowledgment for the service which he had rendered me in recovering my stolen property. I had fully intended to make him a present from the first, but I was, at the same time, determined to resist an attempt to extort one from me. To have given way would have been an encouragement to him to have recourse to similar practices with other travellers. In their interest, as well as my own, I was, therefore, resolved not to yield to his threats.

Suleiman-Ibn-Fais was accompanied by a brother and by some horsemen. The party was joined by one Isaac of Hebron, a Jew pedlar, who traded in small wares with the Arabs of the desert between Kerak and Damascus. He had availed himself of the protection of the Beni Sakk'r chief to visit the encampments of the tribe for the purpose of selling his goods. We assembled at the entrance to the fortress. Ahmed sent a couple of armed men to see me safely through the Aranat Arabs, who might be disposed to revenge themselves for having had to restore to me what they considered their lawful property, fairly acquired after the Arab fashion. I descended the steep declivity on foot. Before crossing a small stream at the bottom of the valley my companions stopped for a few minutes to water their horses and to say their afternoon prayers. As I was making an effort to climb the high pack-saddle of my mule, upon which my carpet and other effects had been placed, I accidentally struck the hammer of one of the barrels of my gun, which went off. The ball with which it was loaded struck a rock hard by, and a splinter from it wounded in the face a deaf and dumb man who was among the sheikh's attendants, and drew blood. He was almost an idiot, and could not be made to understand that this slight wound was the result of an accident. Persuaded that I had purposely fired at him, he drew his sword and made towards me. Had not the sheikh been between us and hastened to seize him, he would probably have cut me down.

During our journey he was constantly watching me with a very sinister expression, and I had to be on my guard against him, as I was convinced, as were my companions, that if he could find the opportunity he would do me some mischief.

We ascended by a very steep and stony path the hills which formed the opposite side of the valley, until we reached a platform upon the same level as Kerak, whose massive and picturesque walls and towers rose in front of us. The country upon which we had now entered was a barren, undulating upland, the western edge of which, from the opposite side of the Dead Sea, presents the appearance of a ridge of mountains, known as those of Moab. It is a great plain which, considerably above the level of the Mediterranean, and overlooking at a great height the Dead Sea and valley of the Jordan, stretches far to the east towards the Euphrates—a desolate wilderness only frequented by wandering Arab tribes, yet with a soil capable of cultivation. During a long ride I saw the ruins of only one village, long deserted. The Arabs called the place Ader. At nightfall we reached an encampment. To my surprise I learnt that the tents were those of Christian Arabs, who, coming occasionally to Kerak, pasture their flocks, which are numerous, in the desert. I was informed that there were four similar encampments of Christians to the north-east of the town. We were hospitably received, two sheep being killed for our entertainment. I spent the night in the tent of the headman, who, finding that I was also a Christian and a European, showed me much civility, and readily gave me the information I asked for about his people. They differed in no way, either in dress or manners, from the Musulman Arabs, with whom they would have been confounded by any one not acquainted with the fact that they were Christians.

I had to sleep, as usual, on my carpet spread upon the ground. The night was bitterly cold. The tent was full of lambs and kids, which, walking and jumping upon me and alighting occasionally on my face, prevented me from sleeping. When I was ready to start I found Sheikh Suleiman-Ibn-Fais quietly drinking his coffee and smoking his pipe, no preparation having been made for our departure. I asked him when we were to resume our journey. He replied that he could not take me to his tents, having business in another direction, unless I was prepared to pay him a sum of money, amounting to about thirty pounds, for escorting me. I refused to do so, and protested against the demand, threatening, if he persisted in it, to return at once to Kerak. He alleged that it was the invariable custom for Frank travellers to pay for the privilege of passing through the territory of the Bedouins when visiting ruins in the desert, and that the son of the Mujelli had assured him that I was a rich Englishman, and was willing to give the sum he required for affording me his protection. In consequence he had actually paid Ahmed a thousand piastres, in order to obtain the usual reward for conducting me to Ammon and Jerash.

I denounced Ahmed as a rogue and unworthy of the name of an Arab, as having failed in his attempt to rob me himself he had sought to induce the sheikh to do so. I declared that I would not, and in fact could not, pay what he asked, and that if he persisted in claiming it I must return to Kerak. Suleiman-Ibn-Fais insisted that at any rate I must repay him the thousand piastres he had given to the Mujelli's son. This I absolutely refused to do, and ordered the man who had accompanied me with the mule and donkey to saddle them and to take me back. But he replied that he had agreed to go with me to the sheikh's tents where he had business, and that he would not, having come so far, return to Kerak. It was evident that he was in the plot for extorting money from me, and that I was entirely in the power of the Bedouin chief to whom Ahmed had, to all intents and purposes, sold me.[3] The Christian Arabs

[3] My readers will remember that Canon Tristram and his party

with whom we had passed the night were unable to assist me against a powerful sheikh of whom they stood in fear; but their chief informed me secretly that I might remain in his tent until an opportunity offered for me to reach Kerak or Jerusalem. He assured me that as long as I was in his encampment I might consider myself in safety.

I accordingly told Suleiman Ibn-Fais that he might recover, in the best way he could, the money of which he had been defrauded by Ahmed, and that as the owner of the mule had refused to return with me to Kerak, I had made up my mind to remain where I was until I could communicate with the Egyptian Governor, or the British Consul at Jerusalem, who would no doubt find means to enable me to return to that place. I offered, at the same time, to pay a fair price for two camels to take me to Jerash, and hence across the Jordan to Tiberias, or any other place where I could be under the protection of the Egyptian authorities.

Seeing that I was determined not to pay the sum that he had demanded of me, he said that, as I was his guest, I might accompany him to his encampment and should enjoy his protection, leaving it to my generosity to give him such recompense

were, many years later (in 1872), the victims of the intrigues and roguery of the then Mujelli of Kerak, who was no other than my friend Ahmed. See *Land of Moab*. Canon Tristram informs me that Ahmed has since been shot in a marauding foray.

as I thought fit. Then, cursing Ahmed very cordially, and denouncing him as 'a dog, the son of a dog,' for having cheated him, he ordered his mare to be brought to him and his attendants to mount. It was nearly midday before we left the tents, the whole morning having been spent in wrangling and disputing with him about money. Although he pressed me hard to pay him what he asked, or at any rate a part of it, he abstained from threats, and was courteous and dignified in his manner. His anger was chiefly directed against the son of the Mujelli, who had cheated him, and upon whom he swore to be revenged.

Our progress was slow, as Antonio and the Jew pedlar were mounted on donkeys, which could not keep pace with the horses. Moreover, Arabs are rarely in a hurry. Early in the afternoon we reached another encampment of Christian Arabs. Suleiman-Ibn-Fais again announced his intention of stopping there for the night. It was useless to remonstrate, and I had to submit to a further loss of time. Two sheep were slain, and in addition to the meat a huge caldron filled with prepared wheat boiled in camels' milk and saturated with rancid butter was placed before us for supper. The tents were pitched near the ruins of a village which the Arabs called Rohetta (?). I observed in the morning that the sheikh no longer wore the handsome silk robe which I had remarked at Kerak. He explained to me, when I asked the reason, that he had given it to Ahmed in addition to the thousand piastres for the privilege of conducting me to the ruins, and he again broke out in curses upon the grasping and cunning youth who had outwitted him in so shameless a manner.

The people of the tents, learning that I was a Christian like themselves and an Englishman, gathered round me in the evening, and I had to answer innumerable questions relating to my country and the state of affairs in Syria. A rumour had already reached them that a renewal of the war was imminent between the Egyptians and the Turks—the latter being supported by England and other Powers—and they were anticipating fresh anarchy on the borders of the desert, of which the powerful Bedouin tribe of the Aneyza would take advantage to attack the weaker tribes on the Syrian confines, and to plunder them of their flocks and herds.

On the following morning, at a little more than an hour's distance from the tents at which we had passed the night, we found another encampment of Christian Arabs. Suleiman-Ibn-Fais dismounted there to breakfast. The chief, an old man, came out to meet him. They embraced each other after the manner of the Arabs, throwing their arms round each other's necks and kissing each other's shoulders. The sheikh, perceiving that I showed surprise, observed that they were old friends, and that although his host was a Christian he was an honest and upright man, which could not be said of some Musulmans—alluding to Ahmed, against whom he again broke out in invectives. We had to stop for nearly three hours, whilst two sheep, which had been dragged to the front of the tent, to be slain in our presence, were cooked for our entertainment. Our track then lay over barren undulating hills until we descended into a deep valley, or gully, called Wady Mojeb, through which a torrent, swollen by recent rains, rolled impetuously. We had to ford it. In the middle of the stream my mule—a weak, emaciated beast—was carried off its legs. It rolled over and I with it. I was entirely immersed in the water, and, with my carpet, and my saddle-bags and their contents, thoroughly drenched. We had to ascend the opposite side of the valley, and about nightfall reached an Arab encampment on a small plateau at a considerable elevation, where we stopped for the night. Near it I discovered a Roman milestone. The inscription was, however, almost entirely effaced. We were evidently on an ancient road which led into the desert in the direction of the distant Euphrates, probably one of the great lines of intercommunication between Syria and the cities on the southern part of that river.

I was wet through, and had no change of clothes. The night was cold, and rain began to fall and soon penetrated through the tent. The Arabs lighted a fire and crouched round it, silent, and glaring at me with their bright glistening eyes. After some time they began whispering to each other, and finally to Suleiman-Ibn-Fais. It was evident that I was the subject of their conversation, and I learnt afterwards that they wished to know who the stranger was, whether I had money with me, and whether they had not a right to a share in it. They were of the tribe of Beni-Hamideh, and arrant robbers. Suleiman-Ibn-Fais congratulated me afterwards on having escaped from their hands, which he attributed to the determination he had expressed to them of protecting me at all risks, and of revenging any outrage that they might commit on a person who was his guest, whether Musulman or Christian. When gazing on their swarthy and ignominious countenances, lighted up by the flaming logs, accustomed as I had of late been to the wild and savage inhabitants of this part of the desert, I thought that I had never before seen such a ferocious set of villains. I was by no means persuaded that I should be allowed to leave their tents with impunity, and what with the cold and my wet clothes and carpet, and with keeping watch the greater part of the night with my gun ready in my hand, I had but little sleep

I was not sorry when daylight appeared and Suleiman-Ibn-Fais gave orders for us to resume our journey. During the previous day he had been constantly pressing me to give him money—at any rate enough to repay him what he had been cheated out of by the son of the Mujelli, and something in addition for the service he was rendering me in protecting me at no small risk from the lawless tribes through which we were passing. He began again upon the same subject almost as soon as we were on our way. If I had no money with me, as I pretended, could I not give him a written order upon the Consul at Jerusalem for, say, 2,000 piastres? He would send one of his own men with it, and I should remain in his tents and visit the ruins until the messenger returned with that sum. I told him that I had no funds with the Consul, and that if a ransom were exacted for my release, Ibrahim Pasha, upon the demand of the British Government, would inevitably have to find it, and would not fail to take measures to obtain its repayment with a considerable addition for himself, so that in the end he, the sheikh, and his tribe would probably be the losers. I advised him, therefore, as a friend, not to press me any further for money which I was unable to pay him, and to get rid, as soon as possible, of a troublesome guest, by accepting what I had offered him for the hire of two of his camels to Tiberias or Souf.

When we stopped early in the day at another encampment of Beni-Hamideh Arabs for breakfast, Suleiman-Ibn-Fais led me to a little distance from the tents, and, out of hearing, continued the discussion about money. He threatened to leave me and to withdraw his protection unless I consented to his demand. As I still refused and a somewhat warm altercation ensued, the dumb man whom I had accidentally wounded when leaving Kerak, and who had been constantly watching me, seeing me return, fancied that I had satisfied the sheikh. He took aside a cunning and brutal-looking fellow who had just arrived—a sheikh of the Salati, a small tribe of very evil repute. By signs he made this man understand that I had a large sum of money in my possession, and that if they could put me out of the way they might share it between them. The sheikh, however, questioned Suleiman-Ibn-Fais as to the property I was alleged to have with me, my appearance and the manner in which I was travelling not tending to confirm what the dumb man had told him. Suleiman-Ibn Fais replied that I was under his protection, and that any money I might have with me of right belonged to him. But the sheikh endeavoured to convince him that he could contrive my death so that no blame should fall upon him, and that they could then divide the spoil. However, Suleiman-Ibn-Fais, although greedy for money like all those Arabs who have been brought into contact with Europeans, had sufficient sense of his duty as a Musulman not to rob or betray a guest who had eaten bread with him. If he could not extract money from me by persuasion, or extort it by threatening to leave me to my own resources amongst people who would plunder and probably murder me without hesitation, he would not use force himself. Moreover, he probably feared lest, having accepted the charge of me from the son of the Mujelli, to whom I had been recommended by the Egyptian authorities, he would be held responsible by them for anything that might befall me.

He came, therefore, to me, and related what had passed between the Salati sheikh and himself. He warned me, at the same time, against the dumb man, who, he said, was a desperate madman and had resolved to have my blood in return for his own which I had shed. Particularly at night, he said, I ought to be on my guard. I replied that

I had been on the watch, and had determined to shoot the fellow if I had any reason to suspect that he was about to attack me. Suleiman-Ibn-Fais gave me to understand that he had no objection whatever to my doing so, but that on the contrary I should be rendering him a service if I freed him from a troublesome follower. He then related to me the history of this man. He belonged to the tribe of Serdé. Having dashed out, with a huge stone, the brains of a cousin who had done something to displease him, he seized a horse, and taking his own brother, a child of two or three years of age, before him, fled to the tents of Suleiman-Ibn-Fais, to whom he attached himself as a kind of attendant. Shortly afterwards the Beni-Sakk'r Arabs had a dispute with a neighbouring tribe. The dumb man, thinking to render a service to his adopted master, managed to enter the tents of those whom he believed to be the sheikh's enemies, and treacherously murdered in cold blood one or two of their inmates. He then fled back to the chief. A blood feud between the two tribes was the consequence. They had fought, and several men had been slain on either side. It was not surprising that the sheikh was anxious to get rid of so inconvenient and dangerous a guest; but, fortunately, I had no occasion to serve him in this respect.

The Beni-Hamideh called the place where they were encamping Agreba. We had to waste several hours there, as the women were obliged to go to some distance to fetch water to boil the sheep which had been killed for us. It was three o'clock before we were again in motion. At sunset we descended into another of the deep valleys or gullies, between lofty precipitous cliffs, through which the waters of Moab are drained into the Jordan. It was called Wady Butum. In it we found a few tents of the Salati Arabs

Next day the ruins of Um-Rasas an ancient city with Christian churches, were visible in the distance, marked by a great solitary tower rising from the undulating plain.[4] I could not persuade Suleiman-Ibn-Fais to deviate from his route in order that I might examine them, and I could not, without exposing myself to some danger, leave the party and go to them alone. Any Arabs that I might meet would assuredly have robbed me, and the sheikh warned me that he would not consider himself responsible for what might happen to me. As all my movements were watched with the greatest suspicion, I thought it prudent to avoid as much as possible making notes and taking observa-

---

[4] These ruins were visited in 1872, and described, by Mr. Tristram. See his *Land of Moab*, p. 141, &c. Um-Rasas means the Mother of Lead or Bullets.

tions with my compass, except when certain that I was alone. If I was seen to write it was immediately said that I was an agent of the King of England, sent to obtain the names of the Arabs that they might be taken as soldiers, and to ascertain where there was water in order that an English army might occupy the country. However absurd these suspicions may appear to be, they were entertained by the ignorant people amongst whom I was travelling, and might have cost me dear.

Again we made a very short day's journey, stopping at every encampment we saw, and winding amongst the low hills without following any direct track or going in any particular direction. I remonstrated with the sheikh. He replied by making fresh attempts to extort money from me. It was evident that he was leading me about the country and purposely avoiding his tents, which could not be far distant, in the hope of getting from me in the end what he wanted. I threatened to leave him and to proceed alone to the ruins of Ammon. But a fresh difficulty then arose. The Arab with the mule and donkey from Kerak declared that he would go no further unless I gave him some money by way of bakshish. Irritated beyond measure by the vexatious proceedings of Suleiman-Ibn-Fais, and losing all patience, I pointed a pistol at the head of the man, menacing to shoot him if he interfered with Antonio, whom I ordered to load the animals. I then drove them before me, leaving the owner to follow us.

We passed during our short day's ride several ruins of ancient towns or villages, which the Arabs said had formerly belonged to the Jews. One of these ruins, on the edge of a deep ravine named Wady Themish, they called Agra Grummal;[5] another, the most extensive which I saw, Moalib. I also observed numerous wells and reservoirs cut in the rock. Although there had recently been heavy rain there was but little water in the latter, and what remained in them was dirty and scarcely drinkable. I was told by the Arabs that in summer they were quite dry; but they must at one time have furnished the only supply of water to the considerable population which once inhabited these now desert uplands. The houses in Moalib, many of which were still well preserved, and one being of considerable size, were constructed of solid stone masonry, and contained low vaulted chambers. The entrances to some of them had stone slabs carved with ornaments.

On the following day we passed a massive dike about fourteen feet thick, built of large dressed stones across a small valley, so as to form a spacious

---

[5] ? The Wady Themed and R'Mail of Canon Tristram's map

reservoir, which was then empty. There was an opening in the centre where there had apparently been a flood-gate or sluice, and on either side of it square outlets, on the sides of which the grooves for raising and lowering gates for regulating the supply of water could be traced. The Arabs named the place Escourt (?) and Sitter (?).

The next ruins we came to were those of Ziza. They consisted of remains of buildings stretching far into the desert, and probably situated on the ancient highway which I had remarked two days before.[6] Soon afterwards we crossed the broad well-trodden track followed by the pilgrims to Mecca. We spent the night in some Arab tents which had been pitched near the remains of a spacious and magnificent building in a style of architecture unknown to me. Ornaments of great delicacy and beauty, carved in the solid stone masonry, covered a part of its façade, decorated the doors and windows, and were carried in bands round the walls. I passed the morning wandering about the ruins, lost in admiration and astonishment at these remains, and speculating as to the origin and history of this marvellous palace thus rising in the midst of the desert. The extreme solitude

---

[6] The ruins of Ziza were visited by Canon Tristram, and are described in his *Land of Moab*, chap. x. He includes the great tank or reservoir which I have described among them.

and desolation of the site was only broken by the occasional appearance of a half-naked Arab, who was curiously watching my movements, persuaded that I was searching for treasure. The suspicion with which I was regarded prevented me from making even a slight sketch of these ruins, which the Arabs called Sheta, or Mashita.[7]

We were now at no great distance from Suleiman-Ibn-Fais' tents. Being at last satisfied that he could get no money from me by further delays, and perceiving that I seemed rather pleased than otherwise by the opportunity which his erratic movements gave me of seeing something of the desert, he thought it better to lose no more time. He learnt from some Arabs, whom we found moving to fresh pastures, where his family and immediate followers had encamped, and made up his mind to join them. After some hours' ride we came in sight of a multitude of camels, and, in the distance, flocks of sheep feeding on the hill-sides. The sheikh recognised his own herdsmen, and a number of Arabs soon gathered round him to welcome him back. Accompanied by a crowd which rapidly

---

[7] In 1872 these ruins were visited and described by Canon Tristram, *Land of Moab*, chap. xi., and the drawings and photographs he made of them enabled the late Mr. Fergusson to identify them as the remains of a palace of the Persian kings of the Sassanian dynasty, whose vast empire extended over Palestine to the Mediterranean Sea.

increased as we went along, we rode to his spacious black tent. As I was to be his guest I alighted at it, whilst the remainder of the party sought their own homes or were received elsewhere.

The following day one of the principal men of the tribe gave a banquet to celebrate Suleiman-Ibn-Fais' return. I was invited to it. Several sheep were killed, and their boiled flesh, hot and steaming, was brought in large caldrons accompanied by the usual wooden bowls containing bread soaked in the gravy, over which sour curds and melted butter had been poured. One of the sheikh's cousins from a neighbouring encampment, with about a hundred armed followers, and as many more Arabs, had been bidden to the feast. The guests squatted on the ground in a semicircle outside the tent. The sheikh and I sat in the place of honour within, and the most delicate morsels were put into a bowl specially reserved for us. After those who had been invited to the feast had satisfied themselves, a promiscuous crowd of men gathered round the caldrons, and when they had eaten, the children were allowed to have the few remaining fragments and to pick the bones. The Arab who had given the entertainment walked, with some of his relatives, round and round the seated groups, superintending the distribution of the viands and inviting his guests to partake of them. Coffee was then served to all, and after they had smoked their pipes they dispersed.

I had now become the guest of the Beni Sakk'r sheikh. He discontinued his attempts to extort money from me, dropped all reference to the subject, and treated me with civility and distinction. His wives made savoury dishes for me. He gave me all the information I asked for concerning himself and his people. Even when he was persecuting me for bakshish and presents, I had found him far more liberal and less prejudiced than other Arabs I had met. He did not appear to entertain any suspicion as to my object in travelling in the desert, helped me in obtaining the names of places, and went out of his way to point out any ruins that he thought might have some interest for me. When I came to arrange with him for the two camels with which he had agreed to furnish me, he even returned a small sum which I had paid him in excess of our agreement. With the exception, therefore, of the annoyance to which I had been subjected during our journey by his unceasing demands for money, I had every reason to be satisfied with him.

The reports which had reached him of the fabulous sums said to have been paid by European travellers for the protection of the sheikhs through whose territories they had to pass when visiting ruins in the desert, had excited that cupidity which seems to be natural to most Arabs. The trick played upon him by the son of the Mujelli of Kerak had naturally roused his anger. Whilst resolved to get back the money of which he had been defrauded, he no doubt thought that he had a better chance of doing so from me than from Ahmed. Having persuaded himself that there was none to be obtained from me, and that I had placed myself entirely in his power relying upon the respect which every true Arab should feel for a guest, the better part of his character showed itself, and he sought to justify the good opinion which I endeavoured to make him understand that I had formed of him.[8]

Suleiman-Ibn-Fais was a Bedouin chief of some power and importance. His encampment was by far the largest I had yet seen. He possessed several well-bred mares, and could command a considerable body of horsemen. In war he was accustomed to wear a suit of chain armour, which

[8] In my subsequent intercourse with the Bedouins I had frequent occasion to observe this double character in the Arab. The same man who at one moment would be grasping, deceitful, treacherous, and cruel, would show himself at another generous, faithful, trustworthy, and humane. The very opposite opinions which travellers, and those who have been brought into contact with the wild independent inhabitants of the desert, have been led to form of them, may be accounted for by this singular mixture of good and bad qualities.

he showed me, and which he informed me had been handed down from generation to generation in his family. Many other sheikhs, he said, possessed similar suits. His gun and his pistols were of English manufacture. He had four wives, and was the owner of several herds of camels and of large flocks of sheep and goats, whose wool and hair formed his principal wealth, being sold to the wandering pedlars who acted as agents for merchants over the Syrian border. He was renowned among the Arabs for his hospitality, and the portion of his tent set aside for guests was always full. Amongst those who were then in it I found a poor youth from Lahore, who had accompanied a caravan of pilgrims going to Mecca from Basra, and who, having tarried behind his companions, had been robbed of his little property by the Bedouins. He was now endeavouring to make his way back on foot to his native country, going from encampment to encampment, and relying entirely for his food upon the hospitality of the Arabs.

I was awoke early one morning by a general movement among the Arabs. They were about to go further into the desert in search of pasture. The greater number of the tents had been already thrown down; the boys were driving away the flocks with loud cries; the women were busy collecting their property, screaming and gesticulating; and the men were placing loads of tent furniture and domestic utensils upon the groaning camels.

As I had no wish to accompany the tribe further into the desert, and was anxious to reach Ammon, which I knew to be in the opposite direction, I begged Suleiman Ibn Fais to let me have the two camels which he had promised to furnish me with. After some time spent in searching for them he returned to say that two camels could not be procured, as they were all required for carrying tents and loads, as the tribe was moving and would probably be on the march for two or three days. But he offered me instead one of his own horses, and a camel for Antonio and my little baggage, and begged me to accept some rice, coffee, and sugar as provisions for my journey. He proposed to accompany me to the tents of a sheikh under whose protection he would place me, and who would conduct me to Ammon and Jerash. We left together the moving mass of human beings, flocks and herds, and took a north-westerly course across some low hills.

We passed during our ride several ruins. Near the remains of an ancient town, which the sheikh called Leban, we came upon a spacious and well-constructed reservoir, which still contained a supply of water and was surrounded by sheep and camels, which had been brought there to drink. At midday we stopped at some tents for breakfast. Taking me aside he informed me that he was compelled to leave me, as he was required to superintend the removal of his tents and to choose a new site for them. He would, however, he said, make arrangements with the sheikh of the encampment, who would furnish me with camels to continue my journey. But no camels were forthcoming, and I had to be satisfied with a wretched horse for myself and a donkey for Antonio. I remonstrated with Suleiman-Ibn-Fais at being thus left by him after he had promised to conduct me to Ammon, but without avail. He assured me that the sheikh under whose protection he had placed me was entirely to be trusted, and it was understood that I was not to be charged for the animals with which he had furnished me, as I had already paid for camels as far as Souf.

He then embraced me after the Arab fashion, and we parted good friends. He mounted his mare and galloped off in the direction of his moving tribe. Accompanied by the sheikh to whom he had confided me, and whose appearance was far from prepossessing, and by Isaac of Hebron, who still believed that under my protection he could reach Damascus, I took the track to Ammon. On our way we passed several ruins, amongst them a large reservoir, on the summit of a hill; an extensive square building of stone masonry, with a cornice and an ornamented gateway; the remains of a small Ionic temple, of which two columns were still standing, surrounded by the shafts of others, and fragments of architraves; and the walls and foundations of various edifices. The name assigned to these ruins was Zug. At a short distance from them I observed some extensive quarries, from which stone for building purposes had been anciently taken.

It was dark before we reached the ruins of Ammon. We could see among them the fires of some Arab tents, which we had, however, some difficulty in reaching, as the ground was encumbered with fallen masonry and rubbish. We were hospitably received, and the customary sheep killed for our entertainment.

CHAPTER IV.

The ruins of Ammon—Sheikh Suleiman Shibli—Fortunate escape—A funeral—The plague—Reach Jerash—Description of the ruins—Isaac of Hebron—The plague at Remtheh—Irbid—A Bashi-Bozuk—Cross the Jordan—Maäd—Deserted by my guide—Tiberias—Hyam, a Jew—His generosity—Safed—The Jew Shimoth—Effects of the earthquake—An Arab muleteer and his wife—Start for Damascus—Robbed by deserters—Kaferhowar—Evading the quarantine—Escape from arrest—Arrival at Damascus.

AT daybreak on the following morning I began my examination of the ruins. They were very extensive, of great interest, and very picturesque, and occupied a long narrow valley enclosed by precipitous rocks, and a second and much smaller valley or ravine leading out of it. Solid walls of dressed stone built across the valley formed the boundaries of the city, which was divided into two parts by a small stream. I was able to make a rough plan of it, indicating the sites of the principal buildings. They consisted of a small temple of a highly florid Corinthian order—the façade of which was almost entire—surmounted by a dome or vault elaborately sculptured in the interior; a building of which the walls, the arched entrance, and a square tower still remained; a second large edifice, with sculptured doorway; a third, still more extensive, on the bank of the stream, with traces of two stories, and containing several vaulted rooms, and part of a portico, with the shafts of four lofty columns; and the great temple, of the Composite order, of which ten fine columns, with cornice and pediment, and many elaborately sculptured ornaments, still remained. The theatre was the most magnificent and best preserved building that I saw at Ammon. It was partly cut out of the side of a cliff, through which some of the vaulted passages leading to different parts of it were carried. It had three rows of seats or gradines, with a recess in the upper row, to which there was access by a square doorway, and which, with two small apsides richly decorated, formed the box or tribune for the president of the games or the Roman governor of the province. Eight columns, a part of the proscenium, were still standing. The height of the whole building, as far as I could ascertain from hasty measurements, was about 112 feet on the outside. On either bank of the stream, which was still crossed by a Roman bridge, were the remains of quays and streets paved with large square flags. Other ruins of temples, monuments, and public and private buildings, among which sheep were feeding, were to be seen on all sides. The city was dominated by a spacious castle, of which the massive stone walls and numerous excavated passages and vaulted chambers still remained. It stood upon the lofty cliff at the junction of the two valleys. Walls and fortifications were also to be traced along the heights around.

The ruins of Ammon could not fail to make a deep impression upon me, both from their extreme beauty and picturesqueness, and from the strange character of the surrounding scenery. At the same time they enabled me to form some conception of the grandeur and might of the Roman empire. That a city so far removed from the capital, and built almost in the desert, should have been adorned with so many splendid monuments—temples, theatres, and public edifices—afforded one of the most striking proofs of the marvellous energy and splendid enterprise of that great people who had subjected the world. Such remains as these show the greatness of Rome, and the influence she exercised wherever she could establish her rule.

It is remarkable that the original names of such cities as Ammon, Jerash, and Baalbec are still retained by the wandering Arabs who encamp among their ruins, although they were occupied and probably rebuilt, and owed all their splendour to the Romans, who gave them the names by which they are best known in history—Philadelphia, Gerasa, and Heliopolis—thus affording a valuable proof of the vitality of traditions in the East.

The sheikh in whose charge I had been placed by Suleiman-Ibn-Fais followed me closely during my exploration of the ruins, and expressed great anxiety that I should bring it to an end as soon as possible. As I perceived several suspicious-looking fellows, who carried firearms, watching my movements, I deemed it prudent to follow his advice. Having spent some hours in examining the remains of the city and in making notes, I returned with him to the tents, and after having eaten continued my journey to Jerash.

We left the narrow valley of Ammon, and entered upon an undulating country, bounded to the east by the range of Gebel Hauran, now covered with snow. The soil appeared to be fertile and capable of cultivation, and I observed here and there green patches of corn and barley, and there were groves of trees in the distance. Peasants, too, were to be occasionally seen driving the plough. I felt as if I were leaving the desert and entering a country with settled inhabitants, and under some kind of government, and that I should no longer have difficulties and dangers to apprehend from the lawless tribes which infest the Syrian borders. I was doomed to disappointment

We had reached the ruins of an ancient reservoir and of a large building near some trees, called by my guide Jahus, when I observed a party of Arabs seated on the grass. Their tufted spears were fixed upright in the ground, and near them were picketed their mares. The sheikh, my companion, recognised the men, and coming to me in great alarm told me that they were a certain Sheikh Suleiman Shibli, of the Adwan Arabs, and some of his followers. They were, he said, notorious robbers, and would certainly not respect the protection which Suleiman-Ibn-Fais had given me. I must make up my mind, therefore, to be plundered of all that I had.

It was too late to retreat. As I had been perceived, it appeared to me that my best course was to advance without showing any signs of distrust. I accordingly went towards the seated Arabs, and giving them the usual salutation, dismounted and took a place in the circle which they formed. My guide seated himself next to the sheikh, who immediately began to question him about me. I could judge by the expression of fear on the countenance of Antonio, who was near enough to overhear what passed between the two, that what the sheikh was saying boded no good to me. When the conversation came to an end I was informed that Suleiman-Ibn-Fais exercised no authority over the tribe to which the country I had entered belonged. The sheikh then demanded the immediate payment of a thousand piastres for permission to proceed on my journey A resistance would have been impossible in the face of some twenty armed men. I endeavoured to diplomatise, and replied that I was under the special protection of the Egyptian authorities, and notably of Suleiman Pasha (the well-known French Colonel Seve), and that I consequently declined to pay what he asked. I added that even were I disposed to do so I had not the money with me, and that he would have to answer to Ibrahim Pasha should any attempt be made to rob me. At the same time I handed to him a letter which had been given to me by Suleiman Pasha for the Mudir of Acre, but which I had not delivered, thinking that he would be unable to read it, and that an official document with a big seal upon it would make some impression upon him.

Unfortunately there was a Mulla of the party, to whom he gave the letter, and who read it out in a loud voice. As it simply recommended me to the good offices of the Mudir, it did not produce the effect that I had expected. The sheikh renewed his demand in peremptory terms, threatening not only to seize my effects, but even 'to cut off my head,' as Antonio translated his menace, unless I at once paid him the money he asked for. He had that mien of mingled suspicion, greed, and cunning which seems peculiar to the Arabs who live on the confines of the Syrian desert and have been corrupted by their intercourse with the Turkish authorities, European travellers, and the village settlers in the adjoining districts. He appeared fully capable of putting his threat into execution, and his followers looked as if they were eager to assist him in doing so.

I repeated that I had no money with me, and I told him that my little baggage was absolutely valueless, as he might satisfy himself by examining it; that to rob and ill-treat me would inevitably get him into serious trouble, as the Egyptian authorities would surely call him to account for any injury that might befall me; and that if he would accompany me to Sheikh Abdu'l-Azeez, whose tents were near Jerash, and for whom I had a letter, I might succeed, by the help of that chief, in coming to some arrangement with him. He asked for the letter, which I gave him. He passed it on to the Mulla, who opened it and read it aloud. When Sheikh Suleiman Shibli heard its contents his manner suddenly changed. He said that it concerned himself, as it was an answer to a request that he had made through Sheikh Abdu'l-Azeez, who was his uncle, to the English Consul at Jerusalem (from whom I had received it) for his intervention in some important matter. He then thanked me for bringing the letter, and said that I might now consider myself under his special protection, and invited me to his tents, which were not far off. I should remain for a day or two his guest, and he would then send an escort with me to Jerash, and even as far as Damascus, if I desired it.

Matters having thus been amicably settled, the sheikh and his followers proceeded to the business which had brought them to the spot. I then perceived that they were seated in a Musulman cemetery. They had come there to bury a dead Arab. The body, which was that of an old man, had been covered with rushes, which were removed. It was then washed and wrapped in a winding-sheet of white linen, the Mulla, who had been brought there for the purpose, repeating the customary prayers and going through the prescribed ceremonies. The corpse was then laid in a shallow grave, and covered with earth and loose stones. A woman, the widow of the dead man, began a dismal moaning and howling, striking her breast and tearing her hair.

I was struck by the appearance of the body, which had a livid aspect, as if the man had met with a violent end. I asked Sheikh Suleiman Shibli the cause of his death. 'The plague,' he answered, ' and he is the third who has died of it in my tents since yesterday' (pointing to two other fresh-made graves). This was said with the careless indifference which is characteristic of Musulman fatalism in the presence of this most dire disease and almost certain death. But I felt that it would be better to hasten away from the polluted spot, and I declined as civilly as I could the invitation to his tents—an invitation which my guide had counselled me not to accept, as my host, he declared, was not to be trusted.

When the burial was over the sheikh took aside Isaac of Hebron, who still followed me on his ass, and whispered something in his ear. Returning to me, he abruptly informed me that the road to Jerash was open, and that I could proceed thither in safety, and that he was unable to give me an escort as he had promised. Although somewhat surprised at this sudden change, I was by no means displeased with it. As he could not furnish me with an escort, I asked him to give me a written document which I could show to any

Arabs of his tribe whom I might chance to meet on the way, and who might be disposed to molest me. After some demur he consented. He directed the Mulla to write a kind of certificate of my being under his protection, to which he affixed his seal. I was not sorry to give him the parting salutation, and to hasten away from the plague-infected party.

Isaac of Hebron explained to me the cause of the sheikh's sudden change of manner. Suleiman Shibli had called him aside to question him as to the property that I had with me. Having satisfied himself that I was really without money, and that my effects were not worth having, he no longer wished to have me as a guest in his tent, and withdrew the offer of an escort, for which he had hoped to be paid, or at least to receive an adequate present. Isaac and my guide told me that I might consider myself fortunate in having escaped from the hands of one of the most notorious robbers and evil-doers in the desert, and expressed their conviction that had I gone to his tents the chances were that I should not have left them again alive. I congratulated myself most on having escaped passing a night in an encampment where the plague was raging.

The greater part of the day had been spent in these discussions, and soon after we had resumed our journey night set in. Although my guide had learnt that an encampment of Arabs was near, he was ignorant of its site. We wandered about, uncertain which way to go. At last we perceived a distant light, and, striking across the rough country, we reached some Arab tents. Before entering them my guide suggested that I should give over my saddle-bags to him. He placed them in a bag of his own, as the owners of the tents, he said, were well-known thieves, and would pilfer anything within their reach. We found a group of wild-looking Arabs gathered round a blazing fire. We dismounted and seated ourselves amongst them, and had no reason to complain of any want of hospitality, for we were welcomed to the fire, which gave an agreeable warmth in the cold, keen air of the desert, and although our arrival had been late and unexpected, a sheep was slain and cooked for us before we lay down to sleep.

I learnt in the morning that the plague had appeared in the encampment, and that it prevailed among all the tribes in this part of the Syrian desert. But what was still worse news, I was told that the Egyptian authorities had established a strict quarantine on the Syrian frontier, and that a line of guards prevented all communication with the country to the west of the Jordan. As I was now only within two hours and a half of Jerash, I determined to proceed there at once, and to decide upon my future plans after I had visited the ruins.

I left the Arab tents before daylight, and early in the morning entered a narrow valley, through which wound a clear stream. Upon its banks, and reaching the steep hills on either side, rose the ruins of Jerash. I was enchanted by the wonderful beauty of the scene, and surprised at the extent and magnificence of the remains. On all sides I saw long avenues of graceful columns leading to temples, theatres, baths, and public edifices, constructed of marble, to which time had given a bright pinkish-yellow tint. Battlemented walls with square towers encircled the city and were carried over the heights above. Outside them were numerous tombs of richly decorated architecture, and sarcophagi which at some remote period had been opened and rifled.

Leaving my Arab boy and Isaac of Hebron to find a night's resting-place for me, I commenced at once an examination of the ruins. They were divided into two parts by a broad street, ending in a triumphal arch at its southern, and a fine gateway at its northern extremity It was paved with large flags, upon which the marks of chariot wheels could still be traced. On either side of it there had been a double row of Ionic and Corinthian columns, of which 153 remained erect. This colonnade opened at one place into an oval of large dimensions, formed by pillars of the Ionic order—probably the Forum; and at others into squares and circles whence streets branched off leading to public buildings. I counted altogether about 250 columns still standing in different parts of the ruins. Innumerable shafts and capitals of others were lying on the ground partly concealed by brushwood.

On either side of this magnificent thoroughfare, which led through the centre of the city from one end to the other, were the great public edifices, and among them several temples. Two of these were of considerable size, and of the rich and profusely decorated Corinthian architecture of the time of the Antonines. The largest of the two had been dedicated to the Sun, as shown by an existing inscription, and stood in an immense double peristyle court.

I found the ruins of two theatres, with their gradines and principal entrances and passages still well preserved. The proscenium of the largest, with its numerous columns and its rich friezes, entablature, and decoration of the most florid Corinthian order, had escaped destruction.

Near the southern entrance was a vast artificial

reservoir which, fed by an aqueduct, supplied the city with water.

I passed the day in examining these interesting and wonderful remains. Forcing my way through the tangled brushwood, I succeeded in making in my note-book a hasty and consequently not very accurate plan of them, which is now of no interest or value except as showing how much of the ancient city still existed at the time of my visit, and how much of it has since been destroyed or has disappeared.

As it was the winter season I saw the valley at an unfavourable time, the trees not being in leaf, and the oleanders and other shrubs not in flower. But I could judge of its natural fertility, and could picture to myself its extreme loveliness in spring, with its multitude of graceful columns and the majestic ruins of the city rising out of a bed of verdure enamelled with flowers, or embedded later in the year in the tinted foliage of a Syrian autumn.

The only tenants of the ruins were a few poor Arab families, some of whom were living in tents, others in the vaulted chambers of the temples or in tombs; others, again, in huts rudely constructed of fragments of ancient buildings, amongst which were exquisite architectural ornaments and mutilated Greek inscriptions. Upon the roof of one of these huts Antonio had spread my carpet, and I endeavoured to seek some rest after my long day's labour among the ruins. But I was soon surrounded by a crowd of Arabs, with forbidding countenances, who gave me no peace. I was not sorry to have as a companion in my night's quarters a Bashi-Bozuk belonging to Ibrahim Pasha's irregular cavalry, who had been sent to the tribes on the borders of the desert to collect some taxes. Had it not been for his presence I should probably have been relieved of the little property that was left to me, as a more ruffianly and truculent set of fellows I had rarely beheld than those among whom I found myself. They had been watching my movements all day, suspecting, as usual, that I was in search of treasure; but fortunately they did not interfere with me.

When I came to discuss with them the means of continuing my journey on the following day, I found them insolent and extortionate. After a long wrangle, during which the Bashi-Bozuk gave me what help he could, I was promised a horse for myself and a donkey for Antonio on the morrow, as far as the village of Remtheh, for which I agreed to pay fifteen piastres. I now discovered that I owed many of my troubles and difficulties to Isaac of Hebron, who had played me false, having been, during the time he had been with me, in league with the Arab sheikhs in their attempts to extort money from me. He was, I found, to have a share in what I paid for the hire of the animals to Remtheh. When I taxed him with his dishonesty he pretended that without his assistance I could not have passed through the desert, nor should I be able to reach the Syrian frontier where I should be under the protection of the Egyptian authorities. He threatened to leave me unless I agreed to pay him a sum of money. I peremptorily refused to give him anything, and bade him go his way as I had no need of his help. On the contrary, I was convinced that it was owing to his having been in my company that he himself had so far travelled in safety. But the sheikh of Jerash, instigated by him, refused to let me have the horse he had agreed to give me if we parted company. I threatened to denounce the Jew to the Governor of Hebron, where his family resided, and where consequently he was well known. As he perceived that the Bashi-Bozuk was observing him with no very friendly eye, and might be a witness against him, he ended by making excuses to me for his conduct.

Wearied by my day's work, and by the angry discussion with Isaac and the Arabs, I retired for the night to a vault in the ruins of the principal theatre, the arena of which the sheikh used as a fold for his sheep and camels. Notwithstanding the high words which had passed between us, and the attempt to cheat me of a few piastres, he killed a kid for my supper, not to be wanting in hospitality. The mixture of grasping avarice and generous liberality in entertaining a guest is, as I have already had occasion to observe, a characteristic trait of even the wild and degenerate Arab tribes which occupy the borders of the Syrian desert.

I was up with the dawn and ready to proceed on my journey, but it was sunrise before the sheikh who was to accompany me could catch his mare, which was grazing among the ruins. We had scarcely left them when we perceived a man who was endeavouring to yoke two refractory oxen to a plough. The sheikh declared that the land to be ploughed belonged to him, and asked that he might help the labourer. Without waiting for a reply he galloped off after one of the oxen which had made its escape. It was a full hour before he returned with the beast. I was vexed and irritated by the delay, as I had a long day's journey before me, and the progress of the donkey which carried Antonio and my little property was very slow. The country was pleasingly diversified with low hills and narrow valleys, and was well wooded with oak, having a park-like appearance. From the higher ground the eye ranged over an extensive and beautiful prospect. In the distance were the snowy summits of Lebanon and Anti-Lebanon; beneath stretched a vast plain, lost to sight in the distance towards Damascus; and to the right were the mountains of the Hauran, also white with snow.

The sheikh pretended that I ran great danger during the day from Arab robbers, who, he said, were wandering about the country in bands, concealing themselves among the hills and waylaying and murdering travellers. However, we saw nothing of them, and the only human being we met was a solitary shepherd tending a flock of sheep. After descending in the evening into a treeless plain and passing one or two small Arab encampments, we arrived at nightfall at Remtheh, a miserable village of ruined huts.

I was compelled to take up my quarters for the night in a room already half-filled with travellers. The rain had begun to fall in torrents, and they, with myself, sought shelter from it in this filthy hovel. But I felt disposed to leave it and the village, and to continue my journey in the darkness and even on foot, when I learnt that the plague was raging in the place, and that some of the inhabitants had died of it on that very day. The sheikh, however, refused to accompany me. I was ignorant of the road, and to have ventured to cross alone and in the darkness this wild and deserted plain, would have been to run no slight risk. I had, therefore, to resign myself to passing the night in the close and infectious atmosphere of the crowded hut.

The intelligence that, owing to the existence of the plague in the Hauran, no one was allowed to enter from the east the neighbouring Pashalic of Damascus, the frontier of which was strictly watched and guarded by patrols of irregular cavalry, caused me the most serious anxiety. Only two days' journey separated me from the city; but the villagers, who congregated in the room when they heard of the arrival of a stranger and a Frank, declared that it would be impossible to reach it by the direct road. Isaac of Hebron, and some pedlars, who like himself were trading with the Arab tribes, held a consultation as to the best means of avoiding the guards by making a detour among the hills. It was suggested that by again striking into the desert and going north for a considerable distance, I might be able to enter the Pashalic by a part of the frontier which was not guarded. But five or six days would be required for the journey. I was so anxious to reach Damascus, where I expected to rejoin my travelling companion, Mr. Mitford, who had now been waiting for me far beyond the time at which I had agreed to meet him there, that I was determined to run the risk of being again stopped rather than remain for an indefinite period in this plague-stricken district.

In the village there were a few Christians. They offered to provide me with two horses, but taking advantage of my difficulties, and greatly exaggerating the risks and dangers of the journey, they demanded the most exorbitant hire for them. They were in no respect better than their Musulman neighbours, from whom they did not differ in appearance, and indeed were more difficult to deal with, and were, if possible, even more grasping, than the Arabs. At length, after prolonged bargaining which lasted into the middle of the night, one of them agreed to provide me with two mules and to take me to Damascus in four days. As I was unwilling to part with the little money that still remained to me, I gave a written promise to pay their hire on my arrival there.

I slept little, as may be supposed. The room was filled with villagers during the greater part of the night. Many of them had been in contact during the day with persons who were dying, or had died, of the plague. Some were perhaps already infected with the fatal disease, and were shortly to be its victims. I had learned that it was making terrible ravages throughout the Hauran, and among the Arab tribes on its borders. I determined therefore to leave the place as soon as I possibly could, and at whatever risk.

Before dawn I was afoot. After some trouble I found the Christian who had agreed to let me have the mules. But he had changed his mind for some reason or another during the night, and refused to fulfil his engagement. There was no one in the village to whom I could appeal to compel him to do so. Although the rule of Ibrahim Pasha nominally extended over the Hauran, there were no official authorities with power to enforce it in this district, which had always been known for its lawlessness. No one seemed disposed to run the risk of falling, with his animals, into the hands of the guards who were watching the frontier, and of being punished with death for an attempt to violate the quarantine.

After much reflection I convinced myself that the plan suggested by the pedlar, to enter the Pashalic of Damascus by a northern route and thus to evade the quarantine, was impracticable; or, at any rate, that, situated as I was, it would be foolish for me to attempt it, and to place myself again in the power of the robber Arab tribes from which I had just had the good fortune to escape. I determined, therefore, to turn southwards again, and to make my way as I best could across the Jordan to Tiberias. But I endeavoured in vain to obtain a horse or mule, or even a donkey, to carry me and my few effects. I could not even find a guide to go with me when I proposed to perform the journey on foot. At last, when I had been driven almost to despair and scarcely knew which way to turn, an Arab offered to let me have two camels as far as Tiberias, asking, however, the exorbitant price of one hundred piastres for the hire of each animal. I was compelled to submit to his demand, and, although the rain was falling in torrents and I had no means of protecting myself from it, I was ready to encounter any discomforts, or indeed any dangers, rather than remain a minute longer than I could help in this hotbed of the plague, especially as the villagers, convinced that all Franks were physicians, were bringing me persons suffering from the disease for whom I was asked to prescribe. The morning was, however, far advanced before the man with the camels was ready. Isaac of Hebron was persuaded that I should fail to reach Tiberias, and that I was running the greatest peril from the Bedouins, who were said to be out on marauding expeditions. He remained behind with his fellow-pedlars, and I saw no more of him.

Owing to the heavy rains that had fallen, the

plain was deep in mud, through which the camels had so much difficulty in making their way that I was compelled to stop for the night at Irbid, a small village less ruined and forbidding than Remtheh, and at no great distance from it. A Bashi-Bozuk, who had been sent there upon some Government business, seeing the deplorable condition in which I was—wet through and not knowing where to go for shelter—very obligingly invited me to share the room which he occupied, gave up the best corner in it to me, and lent me his cloak after I had stripped myself of my wet clothes. He dried them himself at a great wood fire, which was exceedingly welcome to me, as the weather was cold and damp, although I was almost suffocated by the smoke. He completed his hospitality by allowing me to share with him the best dinner he could obtain in the village.[1]

The inhabitants of Irbid were more friendly and more ready to help a traveller in distress than their neighbours. They assembled to see me when they heard that a Frank had come amongst them, as was natural. They offered to furnish me with mules to Tiberias, believing that I should

[1] Although the name of Bashi-Bozuk became afterwards synonymous with everything that was truculent and cruel, I often experienced from these irregular troopers, recruited from all parts and races of the Ottoman Empire, much kindness and help, and found in them amusing and jovial companions.

have no difficulty in reaching the town. I regretted that I had made an agreement with the Arab of Remtheh for his camels, and had paid him their hire beforehand, as he had insisted upon my doing, thus depriving myself of nearly all my money.

The village occupied the site of an ancient town.[2] I discovered in it the remains of a castle, numerous ancient wells, a reservoir, and several houses built of dressed stone and still well preserved. On the lintel of the door of one of them was a wreath in high relief, encircling a Greek inscription, of which only a few letters could be traced.

Irbid had not escaped the plague, but I was assured that there had been but few cases of it in the village. The inhabitants were better dressed, they and their houses more cleanly, than those o Remtheh, which may probably have accounted for their partial immunity from the disease.

The Bashi-Bozuk refused to accept any payment for his hospitality, and would not allow me to depart in the morning until I had partaken of his breakfast. After crossing a deep ravine formed by a torrent, we entered a broad valley leading to the Jordan. We were now in a very fertile and fairly well-cultivated region. On the hill-sides I ob-

[2] ? Arbela.

served two large villages called Zakar—one with a Christian population known as Giaour Zakar, or the Infidel Zakar, to distinguish it from its neighbour—and passed through a third named Kofressed (?). These villages were surrounded by extensive olive plantations. Scattered in all directions were the black tents of wandering Arabs, who, with their flocks, encamp on these hills in the winter and spring. The region to the east of the Jordan is wonderfully rich and fertile, and in the early part of the year produces the finest pasturage.

It was very early in the afternoon when we reached Määd (?), a village on the right bank of the Jordan. My guide refused to go any further, alleging that it would be impossible to reach Tiberias before nightfall, and that we should find no other place to stop at on our way. Although I had made but very little progress I was under the necessity of yielding. The head of the village placed an empty hut at my disposal. But I was not long its sole occupant. I had to share it with some travellers and strangers, flying, they pretended, from a large body of Bedouins who were plundering the neighbouring villages.

Määd was very prettily situated on a green slope descending to the Jordan, which could be seen winding through a fertile valley. In the distance rose the high land and mountains of Palestine. We had crossed the river by an ancient bridge near the ruins of a spacious building which appeared to have been a caravanserai. The stream was broad and rapid. We were, as far as I could judge, about five miles from its outlet in the Lake of Tiberias. Upon its banks were gathered large numbers of herons, cormorants, and other water-fowl. The few villages we had passed had a poverty-stricken appearance, many of the huts being built of reeds.

Ascending next morning a height above the river, we came in view of the Sea of Galilee. From a distance Tiberias, surrounded by a wall with equidistant towers, and standing upon a promontory mirrored in the blue waters of the lake, had the appearance of a great city. The scene was singularly beautiful; but when I reached the town itself I found myself in the midst of a heap of ruins.

We had scarcely caught sight of Tiberias when my guide declared that he could venture no further. He had been told that there were guards forming a 'cordon' round it, who would arrest him as soon as it was known that he came from the other side of the Jordan, would throw him into prison, where he would be kept for an indefinite period, if he ever came out again, and, what appeared to him to be worst of all, that his camels would be confiscated. It was of no use arguing with him or attempting to use force. He proceeded to throw my carpet and saddle-bags on the ground, and to drive his beasts back the way we had come.

There was nothing left to me but to divide my little baggage between Antonio and myself, and to carry it on our shoulders. We were still about three miles from Tiberias, and we had to wade through deep mud. However, we saw no guards nor any sign of the 'cordon.' After a tedious walk we entered the town, which was a mass of fallen houses, through a dilapidated gateway. It was a scene of utter ruin and devastation. I was standing astonished at the sight and perplexed as to where I should find a place to pass the night, when a man in a tall worn-out hat and a threadbare suit of European clothes, and having the long curls on either side of his face which denote the Jew in the East, accosted me in broken Italian or 'Lingua Franca.' He asked me whence I came, and, seeing my forlorn condition and tattered garments, whether I had been robbed by the Arabs. Fearing to tell him that I had come from the other side of the Jordan, lest the Egyptian authorities might learn that I had violated the quarantine regulations, I replied that I was an English traveller from Jerusalem, that I had been robbed on the road, and had been compelled to perform the last part of my journey on foot. I begged him to tell me where I could find a night's lodging, and whether I could make arrangements for continuing my journey to Damascus on the following day.

He appeared to take compassion upon me, and asked me to go with him to his house. He led the way through a maze of fallen buildings, the ruins of which blocked up the streets. Among them were erected wooden sheds for the shelter of that part of the population which had escaped the terrible earthquake that suddenly overthrew the town on New Year's Day, 1837, two years before my visit to it. My obliging conductor inhabited one of these huts, which was divided into several rooms, fitted with divans and a little furniture. He offered me one with a clean European bed—the first I had seen since leaving Jerusalem—and which promised a comfortable night's rest after all the fatigue and privations I had experienced during my wanderings in the desert.

He then informed me that his name was Haym, that he was a native of Poland, and had migrated, like many of his countrymen, to the Holy Land, whence his race had sprung, to spend his last days on the sacred soil. He had been a man of some substance, and having purchased one of the best houses in Tiberias, had established himself as a physician. His stone-built dwelling yielded to the first shock of the earthquake, and in falling overwhelmed himself and his family. His wife and children perished in the ruins. He had miraculously escaped with a broken leg. He was a man of some accomplishments, spoke several languages, and possessed a little knowledge of medicine.

Signor Haym had married his wife's sister, a comely woman, with that fair complexion and light hair which distinguish the Jews in the East from the darker races amongst whom they dwell. She was kindly and hospitable, and welcomed me to their abode. Her husband described to me the earthquake which had reduced Tiberias to a heap of ruins. When it occurred the Jews were for the most part gathered in their synagogues to celebrate a religious festival. They had no time to escape before they were buried under the falling buildings. According to Signor Haym about six hundred Polish Jews had thus perished, besides a considerable part of the Arab population. Four hundred Israelites, he said, still remained in the place, inhabiting the wooden sheds which they had erected on the sites of their ruined houses, and subsisting upon the charity of their friends and co-religionists in Poland and other parts of Europe.

I was anxious to reach Damascus without further delay; but as my host was unable to tell me whether the road was closed on account of the plague, I called upon the Muteselim, or governor —an Egyptian—in the hope of obtaining help and information from him. He was lodged, like the rest of the population, in a wooden shed. It was crowded with his servants, cavasses, and people having business with him. He himself was seated on a low mattress, serving for a divan, at the top of the room. Notwithstanding the shabbiness of my attire, and my not very prepossessing appearance, he received me courteously, and after reading my Turkish Buyuruldi[3] and the papers I had received from the Egyptian authorities, which I had preserved, invited me to be seated, and handed me the long pipe which he was smoking, ejaculating complaisantly, 'Buono! buono!' He fortunately did not question me as to where I had been travelling, but was satisfied with my statement that I came from Jerusalem, had been robbed on the road, and having been deserted near Tiberias by my guide, who had disappeared with his camels, had been compelled to enter the town on

[3] A kind of passport formerly given by the Turkish Government to travellers.

foot. He expressed his sympathy for me, but did not offer to help me in any way, nor could he give me any information as to whether there was a 'cordon' or not between the district of which he was the governor and Damascus. Such ignorance appeared to me surprising even in an Eastern official; but I heard that he was too much occupied in screwing money out of the already impoverished population, over which he brutally tyrannised, to think of other things.

It was necessary to find some means of reaching Damascus. I could not well perform the journey alone and on foot, and I had no money either to hire horses or to engage a guide. I could not conceal the difficulty in which I found myself from my host. With a ready kindness which surprised me he offered to lend me ten pounds, which, he said, I could repay to a friend of his at Damascus on my arrival there. On my observing to him that he was showing an unusual confidence in a person who was a stranger to him, he replied that I was an Englishman, and in distress, and that this was enough. He could, he said, trust to my word; and when I expressed my grateful thanks to him for his generous help, he begged me not to consider myself under any obligation to him, as he had only performed a duty imposed upon him by his religion in succouring a stranger at his gate who was in need.

This noble trait of generosity in a poor Jew made an impression upon me which will, I hope, never be effaced, and has given me a feeling of affection for his race. I could not but contrast it with the mean and sordid disposition of the Eastern Christians.

Assisted by Signor Haym I at last found a Jew who was willing to let me have a horse as far as Safed; but he would not consent to accompany me any further than that town, as he had been told that the roads beyond were infested by robbers, and that if he escaped them he would run the risk of having his animal seized by the Egyptian authorities. My host, however, offered to give me a letter to a brother Jew and fellow-countryman at Safed, who, he had no doubt, would be able to assist me in procuring a mule for the remainder of my journey to Damascus.

As Antonio preferred to return to Jerusalem, and found an opportunity of doing so with a small caravan about to depart from Tiberias, I determined to leave him under the care of Signor Haym, who promised to look after him. The poor boy had served me faithfully through all the dangers and privations to which we had been exposed together. As he was constantly during our journey through the desert in an agony of fear, he must have suffered more mentally than physically, for he was accustomed to hardships. He had fully made up his mind that he would be murdered, and it was pitiable to watch the expression of his face when he was listening to the Arabs, discussing, he declared, whether or not they should rob me and cut my throat.

Having made my arrangements for the morrow I examined, accompanied by my host, the site of the ancient city, which is to the south of the modern town. It was marked by the foundations of ancient buildings, the remains of walls, shafts of columns, dressed stones and fragments of pottery; but there were no ruins of any interest or importance above ground. Near the spot were hot springs over which Ibrahim Pasha had constructed a bath-house—an extensive building which had rather an imposing appearance from a distance, but was already fast falling into decay.

Much refreshed by my night's rest, I left Tiberias early on the following morning, although it rained heavily, resisting Signor Haym's efforts to detain me until the weather had improved. The downpour continued during the whole day. Our path led through deep mud and over swollen streams, until we reached the foot of the mountain near the summit of which Safed stands, when we had to climb over loose stones and slippery rocks. My horse could scarcely make its way through the mire, or find a footing as we mounted the steep ascent. Our progress was consequently slow, and we did not reach the town until after sunset. The hills and surrounding country were enveloped in clouds during the day, and I could only catch occasional glimpses of the lake from the high ground. But there was enough to show me that in more favourable weather the scenery in this part of Galilee must be exceedingly beautiful. The only remarkable thing I observed during my journey was a precipitous rock, forming one side of a deep gorge, in which were a large number of excavated chambers or tombs, similar to those which I had seen in other parts of Syria.

The name of the Jew for whom I had a letter was Shimoth. He received me very hospitably, and invited me to spend the night in his house. Like Signor Haym, he was from Poland, and had migrated to the sacred land of his tribe to die there. He was also living in a wooden hut—Safed, like Tiberias, having been almost entirely destroyed by the great earthquake. His trade was that of a distiller of 'raki,' or ardent spirits, and he was likewise a maker of pipe-bowls. His still was in the shed in which he lived, and was superintended by a sharp intelligent Hebrew boy. Otherwise he was alone, not having wife or family.

I learnt next morning that all communication between Damascus and the south was closed, on account of the plague in the country to the east of Jordan, and that the city was surrounded by a line of irregular troops, who occupied the roads and allowed no traveller to pass. I nevertheless made up my mind to make an attempt to reach that city. Through the help of Shimoth I found an Arab who possessed two mules and who professed himself willing to undertake the journey. Although the rain still continued to fall in torrents and the roads were in consequence almost impassable, I decided upon leaving Safed early on the following day. But when the time for our departure had arrived the Arab absolutely refused to fulfil his engagement, alleging the impossibility of passing through the troops forming the quarantine 'cordon,' who would throw him into prison and confiscate his mules. There was, moreover, he maintained, danger from the Bedouins. I could find no one else in the town who could be induced to hire animals for so perilous a journey. But my host was acquainted with an honest Arab living in the neighbouring village of Zeytun, who also had two mules, and whom he had employed on various occasions in expeditions connected with his trade. He proposed to send for this man and his animals.

In the meanwhile some travellers arrived in the town from Damascus, from whom we learnt that although there was a strict quarantine against all persons coming from the south, yet that it would not probably be enforced against me, if I were provided with a certificate from the Cadi of Safed stating that there had been no cases of plague in the town that I came from, and that I was at the time of my departure in good health. Accompanied by Shimoth, I presented myself to the Cadi, who, learning the object of my visit, referred me to the Muteselim. The governor, an Egyptian officer, taking me for a Jew on account of my being in company with one, refused to grant the required certificate except on the demand and in the presence of the chief dragoman of the Hakhâm-Bashi, or head of the Jewish community. In vain I alleged my English nationality. The man in authority was inexorable, and we had to go in search of the interpreter, whom we at length discovered, and who, on the payment of a small fee, obtained the required document for me.

It was now late in the afternoon. The man with his mules had arrived from Zeytun, and was persuaded to take me to Damascus on my assurance that he would be under my protection, and that if he got into trouble on account of the quarantine I would use my influence with the British Consul to get him out of it. However, as sunset was approaching, he invited me to spend the night in his house in the village, so that we might commence our journey at an early hour on the following morning. To this I readily consented, as I feared that if I remained in Safed further difficulties might arise to interfere with my arrangements.

Although I had lost a day it had not been unprofitably spent. Shimoth being an intelligent man, well acquainted with this part of Syria, gave me a good deal of interesting information with respect to it and the condition of its population. He had been in Safed at the time of the earthquake, which he described in graphic and moving terms—the rumbling underground sound like the roar of distant thunder, the noise of the falling houses, the dust which enveloped the sides of the mountain caused by detached rocks and stones, the cries of the women and children who were buried in the ruins, and the agony and lamentations of those searching for their friends and relations. He believed that about four thousand Jews had perished in Safed alone, the number of victims being greater amongst them than amongst the Christians and Musulmans, as, on account of their religious festival, they were assembled in their synagogues. These buildings, constructed of stone, had buried beneath their ruins almost every soul within them.

The earthquake had been more violent and destructive at Safed than in any other part of Syria. Scarcely a house had remained standing or uninjured, and, as at Tiberias, the inhabitants were at the time of my visit still living in temporary wooden huts among the ruins. There were then, according to my informant, about one thousand Jews left of a population of five thousand, six hundred of whom were of foreign origin, comprising Spanish Hebrews—the descendants of those who, expelled from the Peninsula, had taken refuge in the Turkish dominions—and those who had more recently migrated from Poland and Southern Russia. They were living in great poverty, and most of them in great squalor and dirt. In no part of the world are the poorer Jews distinguished by the cleanliness or neatness of their habits or of their dwellings, which are everywhere to be recognised by a peculiarly offensive odour, caused by the materials used in Hebrew cookery. The Israelites of Safed, like those of Tiberias, unable to gain a livelihood by trade or by any kind of manual labour, were mainly supported by contributions from benevolent Jews in various parts of Europe, conspicuous amongst whom was the beneficent, kind-hearted, and generous Sir Moses Montefiore, who had himself visited them with a view to relieving and comforting them in their sufferings.

I could see but little of the town, which was enveloped in thick clouds during the whole time I remained there. Before the earthquake it had been populous and comparatively flourishing. The houses, built on a declivity, were clustered round the remains of a fine mediæval castle which crowned the summit of the mountain. Beyond the town, and extending to the bottom of the valley beneath, were extensive olive-grounds and vineyards, the produce of which formed the principal articles of trade of the place. Parties of Bedouin horsemen from the east of the Lake of Tiberias had recently appeared in the district, and had plundered a number of villages. Life and property were consequently everywhere insecure, and the general poverty had been increased. The Egyptian Government had promised the sufferers that they should be indemnified for their losses, but the promise had not been kept; nor had sufficient measures been taken to re-establish security on the roads.

Bidding adieu to my host, who refused to accept any remuneration for my entertainment, I descended the hill with my Arab muleteer and reached the village of Zeytun at sunset. One room in his house was already occupied by three Egyptian soldiers, who had been quartered upon him. Another, in which were his wife and family, had a very neat and tidy appearance, the floor being covered with fresh mats. He invited me to take a place in it, and the women at once set to work to make a divan, and to spread carpets in one of the corners. I was surprised at the cleanliness and comfort of the place, which compared very favourably with what I had before experienced in Arab houses. The muleteer himself was an ill-clad and not over-clean little fellow—like men of his calling and class. But his wife was well dressed in a blue silk gown, or rather long loose shirt, and leggings of woollen twist of different colours. She did not think it necessary to conceal her face with a veil. On both sides of her head hung strings of large silver coins, such as were worn by well-to-do peasant women in Syria, and which frequently represented the greater part of their marriage dower. She was tall, erect, and strikingly handsome, with large black eyes, features of singular regularity, and a majestic expression. I thought that I had never seen a more beautiful woman. She gave me an unaffected welcome, and after seeing that the divan on which I was to sleep had been properly prepared, proceeded to superintend the cooking of my supper. Notwithstanding the unexpected neatness and cleanliness of these good people, they were evidently very poor, and the only food they were able to prepare for me and their troublesome and unwelcome Egyptian guests consisted of 'bourghoul,' or dried wheat over which melted butter was poured, and cakes of unleavened bread. When this mess was ready the muleteer's wife carried a large wooden bowl full of it to the three soldiers. She had scarcely entered the adjoining apartment when we were alarmed by her screams, and her husband ran to her assistance. He, too, soon began to call out piteously for help. I rushed into the room, where I found the soldiers belabouring the man and his wife with their 'courbashes.' The presence of a European had a sudden effect upon them. They dropped their whips, and when I said that I should return to Safed to report their conduct to the Egyptian officer in command there, they entreated me in the most abject terms not to complain of them, offering to make any compensation in their power to their victims.

Ibrahim Pasha whilst in Syria maintained strict discipline in his army, and it was his policy to protect and conciliate the population. The soldiers well knew that a representation of their misconduct from a European would entail upon them very severe punishment. Hence their eagerness to induce me to condone their offence. It appeared that, not satisfied with the humble fare which had been placed before them, they had insisted upon having chicken and rice, and on the muleteer's wife representing to them that she was too poor to procure such luxuries, they had set upon her and beaten her most unmercifully, subjecting her husband to the same ill-treatment when he came to her assistance. They both begged me not to complain of the conduct of the soldiers, who, they feared, might revenge themselves upon them when I was no longer there to protect them. Quiet having been restored, I returned to my divan, and received, with the grateful thanks of my hosts, an addition of sour milk and honey to my supper of bourghoul.

I found the muleteer in the morning greatly alarmed by the reports which had reached the village during the night of Bedouins seen on the road to Damascus. He was disposed to shirk his bargain with me. It was only after I had assured his wife that he would be under my protection if any attempt was made to seize his mules and take him for a soldier—a fate which he feared more than being robbed by the Arabs—that he was induced to put the pack-saddles upon his animals. The rain continued to fall in torrents, and the tracks across the country were so deep in mud that we had the greatest difficulty in making our way, being frequently detained for some time before we could find means to flounder through the water-courses. After a wearying ride through an uninteresting country without inhabitants, we descended by a stony mountain pathway to the Jordan, at a spot called by my guide Joseph's Ford. A guard of Bashi-Bozuks were stationed there, in the ruins of a house, and near them some Arabs were living in reed-built huts.

The muleteer declined to go any further that afternoon, as there was no village we could reach before night, and it was dangerous to be out after dark. I was compelled, therefore, to bribe one of the Arabs, who clamoured for 'bakshish,' to allow me to rest for the night in a corner of his hovel, in which, however, I found but little protection from the rain.

The next day we stopped early at the village of Kuneitirah, where we heard alarming rumours of pilgrims going to Jerusalem having been attacked, robbed, and beaten by Bedouins upon the high road from Damascus. My guide determined, therefore, to avoid the beaten track, and to keep in the broken ground at the foot of a range of hills, where he thought we should escape the notice of Arabs on the look-out for travellers.

We had much difficulty in making our way over the rocky ground, and across the innumerable watercourses swollen by the rain, which had now been falling incessantly for several days. We were by the side of one of these torrents, seeking for a place to cross it, when we were suddenly surrounded by a number of men armed with guns. They were not Bedouins, but from their dress evidently conscripts who had deserted from the Egyptian army. I had left my mule, and one of these men, with a companion, began to turn out the few articles that still remained in my saddle-bags. The others seized me, and demanded money, threatening to shoot me if I refused to give it to them. Resistance was useless. I offered them some loose piastres I had in my pocket, but this did not satisfy them. They compelled me to take off a part of my clothes, and perceiving round my waist a wash-leather belt in which I carried a few gold coins, they tore it off me by force. They then asked for tobacco, and made me give them what I had. After examining the contents of my saddle-bags, and taking a few articles of no value, and allowing me to keep my gun, which was of no use to them, my books, papers, compass, and medicines, they went off, carrying with them a part of my clothes, and leaving me in my trousers and shirt, and with my Arab cloak, which was now almost in tatters and not worth taking.

As soon as they were out of sight the muleteer, who had taken to his heels and had hidden himself as soon as the robbers appeared, returned to me. He was overjoyed to find that his mules had not been stolen. The deserters, who were hiding themselves from the Egyptian authorities, had no doubt thought that the animals would be in their way. We collected the few things left to me, which were scattered over the muddy ground, and made the best of our way across country to a village called Kaferhowar (?).

Here we learnt that several parties of deserters from Ibrahim Pasha's army such as we had met were robbing travellers and plundering villages. The conscription had been introduced for the first time among the sedentary Arabs who inhabit the eastern borders of Syria. It was enforced with great severity and cruelty, and to avoid it many villages had been deserted by their inhabitants. The conscripts, who were led off as prisoners to Damascus, took the first opportunity to escape and to return to their homes, or to conceal themselves in the mountains and in the desert, infesting the highways and despoiling single travellers and even caravans. They were able to commit their depredations with impunity, on account of the quarantine, which prevented the villagers from entering Damascus to lay their complaints before the Egyptian governor, Sherif Pasha. We were informed at Kaferhowar that the 'cordon' was maintained very strictly, and that at no great distance from the village we should fall in with patrols of irregular cavalry, who would turn us back or probably fire upon us if they suspected we had the intention of violating it.

My muleteer was so much alarmed by what he heard from his friends in the village as to the punishment which awaited him if he attempted to evade the quarantine—the seizure of himself for a 'nizam' (regular soldier), and the confiscation of his mules—that he absolutely refused to go any further, offering to forego the hire for his animals rather than run the risk of losing them altogether, and of finding himself, against his will, a soldier. I had no money to pay him, having been robbed of the little I had about me in the morning; I could only do so on my arrival at Damascus. From a height near the village the city could be distinguished in the distance, its gardens forming a dark line on the horizon. To be so near it and not to be able to reach it, without money and almost without clothes, and not knowing where to go, I was well-nigh in despair. My guide, who was at bottom a good fellow, and appeared really to feel for me in the disagreeable and somewhat hopeless position in which I was placed, did his best to help me. After consulting with the sheikh of the village, he informed me that he had found a man who would take me to Damascus, avoiding the quarantine, if I agreed to pay him a small sum on my arrival there. But I should have to travel during the night on foot, and to follow his directions in everything. I made up my mind to run any risk rather than remain in my helpless condition, and at once agreed to the terms proposed. I promised further to send back to Kaferhowar by my guide the money I had to pay to the muleteer for the hire of his animals.

We were to leave the village in the evening. The rain was still falling heavily, and the night promised to be very dark. This was all in our favour, as we should, we hoped, be able to conceal ourselves from the patrols.

Kaferhowar is divided into two parts by a small stream; the part on the northern side was called Beyt-el-ma (the house of water). I discovered in the village the remains of ancient buildings, marble slabs, the shafts of columns, and the basement of a temple constructed of blocks of white marble, which the inhabitants said had been built by Nimrod! This was the first time during my wanderings in the East that I found the name of this mythical personage associated with an ancient monument. I had no time to search for inscriptions.

Soon after sunset, the Arab, whose name was Ahmed Saleh, having shouldered my carpet and divided with me my little luggage, we left the village together, and entered upon hilly and broken ground. It soon became pitch dark. My shoes were almost worn out, and as we had to walk on loose stones, and to climb over stone walls, I suffered much inconvenience and pain, and soon became sore-footed. It was evidently not the first time that my guide had evaded the quarantine. Although we could not see a yard before us, and the rain continued to descend in torrents, he went steadily on his way, wading through swollen rivulets and deep mud, scrambling over rocks, and creeping through ditches and watercourses. I followed him silently, making as little noise as possible. We walked for some hours, occasionally stopping for a few minutes, as I was nearly exhausted. When the day broke we could see the gardens of Damascus within a short distance of us. Ahmed Saleh assured me that we had passed through all the patrols, and that we might now consider ourselves in safety. We sat down to rest before entering upon the broad beaten track which led through the forest of fruit and other trees surrounding the city. We had scarcely resumed our walk when we perceived a horseman galloping towards us. He proved to be a Bashi-Bozuk.

He came up to us and inquired whence we came. Our answer not satisfying him, he ordered us to turn back and to accompany him to the officer in command. I had preserved the certificate given to me by the Muteselim of Safed, and I handed it to him, stating that it was a permission from a competent authority to enter Damascus, and that he had consequently no right to stop us. After he had looked at the document, which he could not read, and had examined the seal, he returned it to me, saying that it must be shown to his chief. Fortunately I had found at the bottom of my saddle-bags a small gold Turkish coin, which I slipped into his hand. It produced more effect than the paper. He looked at it for a moment, and after a little hesitation put it into his pouch and left us.

I hurried onwards as fast as my weary legs could carry me to the gardens. But we had scarcely reached them when we perceived the Bashi-Bozuk again galloping after us. He soon overtook us, and holding out the coin said that it was a bad one and asked me to change it. When I told him that I had no other to give him, he ordered us to go back with him, saying that he would certainly be shot when it was known that he had allowed us to violate the quarantine, and that it was his duty to take us to his officer. I replied that I was ready to accompany him, but I warned him that I should denounce him to his chief and to the authorities at Damascus for having accepted a bribe, as he had only wanted to return the gold coin when he had doubts as to its genuineness, and that I should make such representations, through the English Consul, as would insure his condign punishment. Seeing that I was willing to turn back, and reflecting no doubt that the complaints of a European through a Consul might get him into serious trouble, he thought better of the matter, looked again at the piece of gold, assured himself that it was really worth five piastres, and then retraced his steps, leaving us at liberty to proceed. We lost no time in doing so, my guide leaving the high road to avoid further observation, diverging into by-lanes and climbing over the ruined walls of gardens.

Overjoyed at having thus escaped from the horrors of a quarantine of perhaps forty days in a filthy Arab hut, I almost ran until we were within the gates of the city. We passed through them with a crowd of peasants bringing their produce to market—the guards, no doubt, taking us for poor people from a neighbouring village. It was late in the morning before we reached the British Consulate, through numberless narrow winding streets enclosed by the naked walls of mud-built houses.[4]

[4] When, as the Queen's Ambassador to the Sultan, I entered Damascus in 1878, my thoughts could not but revert to my entry into the same city nearly forty years before. The contrast was singular enough. On the second occasion I experienced a reception such as, I believe, had never been accorded to any European, whatever may have been his rank, in the Turkish dominions. Midhat Pasha, the Governor, and all the authorities, Musulman and Christian, came out several miles to meet me. Abd-el-Kadr, the celebrated chief of the Arabs of Algiers, then an exile, received me on the way in his country house with a sumptuous entertainment. As we drew near to the city we passed through vast crowds of men and women of all creeds—Mohammedans, Christians, Jews, &c.—with their respective chiefs, who had come out to welcome me. It was a sight never to be forgotten. Similar demonstrations awaited me in all the towns and villages through which I passed during my tour in Syria and Palestine.

CHAPTER V.

Mr. Consul Wherry—A Turkish bath—Damascus described—An Arab barber-surgeon—Padre Tomaso—Persecution of the Jews—The French Consul—Purchase a mare—Leave Damascus—Cross Anti-Lebanon—The Mutuali—Arrive at Baalbek—An Italian military instructor—The Emir—Meet with an accident—Leave for Beyrout—Cross Lebanon—Reach Beyrout—Journey to Aleppo—Rejoin Mr. Mitford—Leave Aleppo for Baghdad.

MR. WHERRY was then British Consul at Damascus—a courteous and well-informed gentleman, who had long held similar offices in the Levant, and was one of those honourable and useful public servants in the East who have been very unjustly and foolishly stigmatised as 'Levantines.' He was not a little surprised at being addressed by an Englishman clad in scarcely more than a tattered cloak, almost shoeless, and bronzed and begrimed by long exposure to sun and weather and to the dirt of Arab tents. I made myself known to him. He was expecting me, as Mr. Mitford, after long waiting for me, had gone to Aleppo, leaving a letter for me. My fellow-traveller wrote that as the time at which I had promised to meet him at Damascus had long gone by, and as he had not heard from me, he had made up his mind to continue his journey to Baghdad, but finding that the direct road thither through the desert was impracticable on account of the troubled state of the Bedouin tribes, he had gone to Aleppo, where I might still be able to rejoin him.

Mr. Wherry, seeing my exhausted condition—for I could scarcely stand after the fatigue I had gone through during the night—kindly offered me some tea. I had not tasted tea for many weeks, and it would be difficult to describe how delicious I found it. After I had rested a little he sent one of his janissaries[1] with me to the Latin Convent, where the Friars gave me a room, very barely furnished, but which appeared to me to contain every luxury, after what I had been of late accustomed to. My first thought was to take a bath, and to provide myself with some clothes. I went to one of the principal Hamams of the city, and was somewhat surprised that in my ragged condition I obtained admission. But in those days with good Musulmans there was no distinction of persons, and the principles of equality were not only professed but practised by them. I found when

---

[1] The guards appointed by the Turkish and Egyptian authorities and attached to a Consulate for the service and protection of the Consul, now called cawasses, were then termed janissaries.

travelling in the East and undergoing great fatigue, there was nothing so refreshing as the Turkish bath. After I had gone through the various processes—had been soaped and kneaded, and had my joints pulled and cracked—I smoked a narguilé, and fell into a sound sleep on the divan with its clean white linen, on which the bathers found repose before dressing. In the meanwhile I had sent the janissary to the bazar to buy me some ready-made clothes. A European dress was not to be obtained, and I had to be satisfied with that worn by the Egyptian Nizam, or regular troops, in Syria during winter—a pair of baggy trousers with tight leggings, a short jacket, a waistcoat fastened up in front with numerous buttons, a coloured sash of common English materials, and a linen shirt. Thus clothed, and having given my discarded garments to one of the attendants at the bath, to be thrown away, I returned to the convent.

Damascus has been so often described that I need scarcely write anything about the city—its narrow streets, deep in filth and dust or mud, according to the season of the year, and its houses with exterior walls of earth, without windows or architectural decorations of any kind, but enclosing spacious and beautiful courts with fountains of ever-running water, orange trees, and beds of flowers, into which open rooms adorned with the most exquisite carvings and with designs in gold and the brightest colours. When the traveller, after passing through the long covered entrance which led into these apparently half-ruined and ignoble dwellings, suddenly found himself, as if by enchantment, in the midst of one of these luxurious and beautiful edifices, he might fancy himself in a palace described in the 'Arabian Nights.' In such a house lived the English Consul.

At the time of my visit the city was full of Egyptian troops, and had a busy and prosperous appearance. The extensive bazars were crowded with men and women of many races, and in endless varieties of costume—Egyptian soldiers, Christians and Musulmans from the surrounding villages, Maronites from Mount Lebanon, Druses from the Hauran, Bedouins from the desert, and inhabitants of Damascus itself in their gay robes of silk and ample turbans. The East had not then experienced the change that contact with the West has since brought about, and the dress, manners, and habits of the people of Syria were still what they had been for many generations before.

I spent much of my time during the few days that I remained in Damascus in the bazars, enjoying the lively and picturesque scene. The shops were then filled with rare and beautiful silk manufactures, with quaint furniture, inlaid arms, and a thousand curious objects for which the traveller would now search in vain. I used to sit in the shop of a barber, with whom I had made acquaintance in the following manner. When I arrived at Damascus I was suffering excruciating pain from a whitlow under one of my thumb-nails. Not knowing where to go to obtain relief, I entered a barber's shop in the bazar, thinking that the owner probably followed the trade of surgery as well as his own, like his brethren in other parts of the world. I showed him my thumb. He was a tall, muscular fellow, and grasped it with a grip of iron. He then took a sharp instrument, and, inserting it under the nail, drove it into the sore. In vain I struggled and howled, as the agony I experienced was intense. He held me as if I had been an infant, until he had pressed the matter out of the opened whitlow. He then allowed me to withdraw my hand, and turned with a look of satisfaction to the little crowd which had gathered round his shop to witness the operation.

I went to him daily to have my finger dressed with an ointment which he prepared. The cure was complete, but the method, to say the least of it, was somewhat brutal, and I vowed that, after my experience of Arab dentistry and surgery, I would not again trust myself to a Bedouin to draw a tooth, or to a Damascus barber to cure a whitlow.

The city was at that time in great commotion, on account of the disappearance of an aged Franciscan friar—an Italian, known as the Padre Tommaso—and his servant. The general belief was that he had been murdered by the Jews for his blood, to be used in making certain cakes for the feast of the Passover — a traditional accusation against this much-persecuted and calumniated people, which has existed from time immemorial in the East, and, unhappily for them, in the West also, only the victims of this horrible sacrifice have usually been supposed to be tender children, and not tough old monks.

However, Padre Tommaso, being a Roman Catholic priest, was under French protection, and the French Consul, the Comte de Rattimenton, had taken up the case with the greatest energy and even passion. He appeared to have assumed that the popular accusation against the Jews was well-founded, and he proceeded to visit and search the houses in the Jewish quarter, accompanied by the janissaries of the Consulate and guards furnished by the Pasha.

Padre Tommaso practised as a physician, and among the patients he visited were Jews as well as Musulmans and Christians. It was said that on the day he had disappeared he had left his convent for the Jewish quarter, and had not been seen since. He was attending a wealthy Hebrew merchant who was ill. It was suspected that he had gone to this person's house, which was consequently subjected to a most rigorous search by the French Consul personally attended by his guards. A barber, upon whom suspicion had fallen, was arrested and so cruelly tortured that he was forced to state that he had been called to the house of the merchant, and had there been employed to put the padre to death. The blood of the murdered man, he was further made to declare, had been collected in earthen vessels, and the body, after having been hacked into pieces, buried in a cellar.

After two or three days, during which the Jewish merchant's house was examined in every part, a few bones were discovered in a drain. They were pronounced by some Italians, utterly ignorant of anatomy or of medicine, who served as surgeons in the Egyptian army, to be those of a human being; others maintained that they were those of an animal. The Comte de Rattimenton, nevertheless, found in them enough to justify a suspicion that they might be those of the missing friar. The male members of the family, which was one of the most respected and wealthy in Damascus, and many Jews who resided in their neighbourhood, were thrown into the common jail. The ladies were insulted, ill-treated, and confined to their apartments as prisoners. The French Consul, who had superintended these shameful proceedings, finding that he could obtain no evidence to prove that the persons arrested were guilty of the crime attributed to them, called upon the governor of the city to use torture to extract a confession from them. Sherif Pasha, who held that office, had the reputation of being a just and humane man, but he found himself compelled to yield to the Consul's threats. The unfortunate Hebrew prisoners, amongst whom were old and infirm men, were subjected to the bastinado, and to other cruel treatment of a shocking description. But no evidence was elicited from them to inculpate the accused. These revolting proceedings continued after I had left Damascus, and until public opinion in England and France indignantly denounced a persecution worthy of the Middle Ages.

The French Government, ashamed of the part played in the matter by its Consul, removed him to another post. The unhappy Jews were released, after having undergone terrible sufferings. Subscriptions were raised for them in England and elsewhere, and the benevolent and generous Sir Moses Montefiore, taking up with his usual devotion and warmth the cause of his persecuted co-religionists, came himself to Syria to examine into the matter, and afterwards went to Constantinople, where he obtained a firman from the Sultan which insured to his Jewish subjects protection from similar outrages in the future. The fate of Padre Tommaso, however, remained a mystery. He had probably been enticed into some house, and had been murdered for the sake of the little money that he usually carried about him. The popular tradition at Damascus may still be that he was slain by the Jews for his blood, to be used for making their sacrificial Passover cakes.

I was desirous of losing as little time as possible, and of going to Aleppo by the direct road through Hamah and Homs; but as by a short détour I could see Baalbek, I resolved upon visiting those celebrated ruins on my way. I had been able to obtain a little money from the Consul on my letter of credit, and had paid what I owed the generous Jew of Tiberias, as he had requested me, to his agent. I had also sent the hire of his animals to the muleteer whom I had left at Kaferhowar. My means were so small, and the journey before me so long, that I was compelled to travel with the utmost economy. Not being able to find a man who would let me have only one mule and accompany me on foot to Aleppo, I determined to buy a horse and to perform the journey as I best could alone. I should thus be entirely independent, and be able to follow the route which suited me best.

Accordingly, early one morning I went to the Damascus horse-market, held in a 'maidan,' or open space, in which a miscellaneous crowd, chiefly consisting of Bashi-Bozuks and Arabs from the villages, were buying and selling horses, mules, and donkeys, quarrelling and screeching at the top of their voices. In the din and confusion it was not easy to make out the proceedings. The owners of animals for sale were leading them about, accompanied by an official crier, who called out the sum last offered and invited the bystanders to bid. I observed a strong well-built mare belonging to an Arab, which was being shown about in this way.

She appeared to me to be just the kind of animal I wanted. When the crier came near me I advanced a few piastres upon the last bid, and, after a little delay, he returned and informed me that the mare was mine for about 10*l.*, and demanded his customary fee upon the transaction. The price appeared to me very reasonable, and I was well satisfied with my bargain. I hired a boy to lead the mare to the convent, and on my way through the bazar bought a native saddle, over which I could throw my carpet and my saddle-bags.

I left Damascus accompanied by a Bashi-Bozuk, who, Sherif Pasha had informed the British Consul, was proceeding on business to Baalbek, and would act as my escort on the way. He was furnished with a Government order directing the sheikhs of the villages at which we might stop to provide me with food and lodging, and provender for my horse, at the Government expense. I soon had an opportunity of learning how well-founded were the complaints against the Bashi-Bozuks, who then overran the country, and were employed by the Egyptian authorities in collecting the taxes and tithes, in guarding the roads, and on other business. I had constantly to interfere to prevent my companion from ill-treating the sheikhs and inhabitants of the villages through which we passed, if they did not bring at once what he pretended to consider necessary for my entertainment, but which was really for his own. His 'courbash' and the butt-end of his gun were in constant requisition. The only answer that I could obtain to my remonstrances was, when the inhabitants were Christians, that they were pigs and had to be driven by the stick, and when Musulmans, that they were asses who could only be treated in the same way.

As my Bashi-Bozuk was in no hurry, and by the aid of my firman was living upon the fat of the land, he insisted upon stopping at almost every village on our way. Having been obliged by my companion to stop for the night even before we had left the gardens surrounding Damascus, it was not until the day after I quitted the city that I ascended the Anti-Lebanon. The beautiful view obtained from its summit over the city, with its gardens, its minarets, and its running waters, and the boundless desert beyond, was soon shut out from me by a dense snowstorm, through which we could scarcely make our way. After struggling against it for some time we were forced to turn back and to take refuge in the Christian village of Dimas, halfway up the mountain.

We had considerable difficulty in crossing the pass on the following day, on account of the snow which had fallen in the night. We descended the western declivity of the Anti-Lebanon range to a village called Zibdani, in a valley watered by a stream of the same name. Here we learnt that the Mutualis, a fanatical and lawless tribe, to whom were attributed strange idolatrous rites, through whose country we were now passing, were in open rebellion against the Egyptian authorities, in consequence of an attempt to enforce the conscription among them. The sheikh of the village, a venerable old man with an ample white beard, who belonged to the Mutuali sect, tried to persuade me not to attempt to reach Baalbek without a strong military guard. He warned me that the road was very insecure, and that I should run great danger of being robbed, and, if taken for an Egyptian functionary, as would probably be the case, of being murdered. The Bashi-Bozuk took alarm at what he had heard, and refused to incur the responsibility of allowing me to proceed without a sufficient escort to insure my safety. As I saw that he was determined not to go any further, I resolved to continue my journey alone. The sheikh, finding that I could not be prevailed upon to turn back, expressed his opinion that I should be safer without the Bashi-Bozuk than with him, as the Mutualis bore no ill-will to Europeans, whilst they would certainly cut the throat of any Egyptian who might fall into their hands. He offered to send with me to the next village one of his followers, who would explain that I was an English traveller to any of his people we might fall in with on our way. I started with this man, leaving behind me the Bashi-Bozuk, who feared that he would be punished for having deserted me on the road should Sherif Pasha learn that he had allowed me to proceed alone. To calm his fears I gave him a certificate in writing that he had left me at my own request.

I met with no adventure, and reached in the

afternoon a large village, in which I was very hospitably entertained by a Turk employed on some Government business, who took me for a European physician belonging to Ibrahim Pasha's army, as I wore the Nizam dress. I was on the beaten track to Baalbek, and had no difficulty in reaching the place early in the following afternoon. The Mutualis did not molest me, and although I passed through several of their villages I saw no signs of the insurrection, the reports of which had, no doubt, as usual been greatly exaggerated. As I approached the ruins I could see the stately remains of the great temple rising above a collection of low flat-roofed mud cottages. I found my way to the residence of the governor of the place, who was a Mutuali, and a member of the family which possessed the hereditary chieftainship of the semi-independent clan occupying the broad valley, or rather plain, of Baalbek. He was still styled the 'Emir' (Prince), and had been recognised and maintained in his authority by Ibrahim Pasha. He received me civilly, surrounded by a number of armed followers. Learning that I was a European, he offered to conduct me to a Frank, who was living, he said, in the village, and could speak my language. He then took me to a house occupied by a Syrian Christian from Damascus, who was employed as a tax-gatherer, and who had the reputation of being well acquainted with Italian. However, his knowledge of that tongue was confined to *favorisca*, which he kept constantly repeating, and one or two other words. The Emir left me with him, and I obeyed his '*favorisca*' by sitting down beside him on his divan. He then directed some native 'raki,' or brandy, to be brought to me, thinking, no doubt, that it was the first and most urgent requirement of a Christian.

Finding that his knowledge of Italian was too limited to enable us to exchange many ideas in that language, he proposed to take me to the house of a European who was quartered in the village with a squadron of Ibrahim Pasha's cuirassiers, which he was engaged in instructing in cavalry drill and manœuvres. Signor Ferrari, the gentleman in question, was a Neapolitan. He received me very courteously, was delighted to find some one with whom he could converse in his own tongue, and insisted upon my accepting his hospitality so long as I remained in Baalbek.

The evening was drawing near before I had settled myself for the night in the house of the obliging cavalry instructor. Having done so, and stabled my mare, I paid a hasty visit to the magnificent ruins, which, lighted up by the setting sun, rose high above the mean and squalid dwellings clustered around them. I was lost in admiration and astonishment at their stupendous proportions and their marvellous beauty. The stately columns and the blocks of richly sculptured marble which still kept their places in the great temple had assumed that exquisite golden hue which I had observed in the remains of Ammon and Jerash. I had no time to examine in detail all the wonderful monuments by which I was surrounded. I could only go from one to another, and then linger among them until they were clothed in darkness.

Early on the following morning I called on the Emir to ascertain from him the state of the country between Baalbek and Aleppo, concerning which very alarming reports prevailed. When he learned that I had the intention of going alone to Homs, he emphatically declared that he would not allow me to proceed in that direction without an escort of at least twenty Bashi-Bozuks and twenty Mutuali villagers. The whole of the Mutuali tribes to the north of Baalbek were, he said, in insurrection against the Egyptian Government; his own brother, who had attempted to restore order, had been killed, and if anything happened to me he would be held responsible. I was to give him, moreover, a written declaration that I considered the escort which he proposed to send with me sufficient for my security, and that I had taken the road to Homs of my own free will after having been warned by him of its danger.

Finding that he was determined not to allow me to follow the direct route to Aleppo without this large escort, for which I could not afford to pay, and which would, I felt convinced, take to flight at the first appearance of any real danger, I very reluctantly renounced my intention of going to Aleppo by the way of Homs, and decided upon taking the more circuitous route by Beyrout and Tripoli. This would cause me a delay of several days, which I could ill spare, but there was no help for it.

The Emir having promised to send a Bashi-Bozuk on the following morning to accompany me to Beyrout, the road in that direction being, he said, perfectly safe, I proceeded to examine the ruins at my leisure. I spent the whole of the day among them, delighted beyond expression with their beauty and splendour, and more impressed than ever with the culture, energy, and power of that wonderful people which had planted their colonies in the most distant lands, and had adorned them with such magnificent and unrivalled public works.[2]

---

[2] The ruins of Baalbek have been so frequently described since my visit to them that it is needless for me to record more than the general impression which they made upon me. In the year 1879 I

In the evening Signor Ferrari lent me a horse, and took me to see some ruins in the neighbourhood, and the enormous blocks of dressed stone sixty feet in length, which a people far more ancient than the Romans had prepared for the foundations of their great temple to Baal, and which still remained near the quarry whence they were hewn. As we returned from our ride an accident which might have proved very serious, if not fatal, befell me. My friend's horse was a spirited and ill-trained animal. It took fright and ran away with me. My saddle-girths broke, and I was thrown with great force upon rocky ground. I was taken up stunned and insensible. Under the care of my host, who had me carried to his house, I soon recovered my senses. With the exception of a few bruises I was none the worse for my fall, and after a good night's rest felt

spent two days among them to examine them at leisure, being furnished with every convenience, and indeed luxury, by Midhat Pasha, in whose jurisdiction as Governor of the Province of Damascus they then were. I was accompanied on my way to Baalbek by his secretary and a number of Mutualis, with their principal chiefs. A large party of horsemen from the great Bedouin tribe of the Aneyza, with their sheikhs, met me on the road and escorted me to the small town which has arisen among the ruins. I was grieved to find that since my previous visit many of the delicate architectural ornaments of the temple and other buildings had greatly suffered from wanton injury attributed to English and American tourists. I urged Midhat Pasha to take measures for their preservation.

myself sufficiently recovered to pursue my journey on the following morning.

The Bashi-Bozuk who was to accompany me did not, however, make his appearance. It was only after several personal applications to the Emir, who seemed disposed to detain me, as a report had reached him that four Egyptian soldiers had been murdered that morning in the neighbouring village of Fica, on the Aleppo road, that I was able to procure my guard. It was already midday before he was ready. We left Baalbek in a violent storm of snow and hail, and I was half-blinded by it as I crossed the treeless plain, then deep in mud, which separates the Anti-Lebanon from the great Lebanon range. I was glad to reach, about sunset, the village of Malaga, where I hoped to find shelter for the night. Its inhabitants were Christians. The headman was a surly, inhospitable fellow, who refused to give me any help or to procure me a lodging. After in vain trying, by offers of payment, and by threats of complaining to the Egyptian authorities, to induce him to find me a room and something to eat, I was compelled to take refuge in a kind of barn, half filled with barley and straw, which fortunately furnished food for my mare, and to lie down to sleep supperless and shivering from the cold.

During my dispute with the sheikh the Bashi-

Bozuk, whose duty it was to find lodgings and food for me, had disappeared. He did not show himself again until the following morning, having, no doubt, settled himself somewhere comfortably for the night, leaving me to my fate. When I reproached him for his conduct he became impertinent, and used threats to extort money from me. Finding that he could not intimidate me, he said that his orders were to accompany me to Malaga, and not further, and that he would return to Baalbek. I begged him to do so, as I had no need of his services, and, mounting my mare, took the road to the large village, or rather town, of Zahlé, which was scarcely more than a mile distant. It was inhabited entirely by Maronite Christians under the rule of the Emir Beshire, then the great Druse chief of the Lebanon, and had a flourishing appearance when compared with the surrounding Musulman and Mutuali villages. The place was deep in snow, and I had some difficulty in reaching it. The Muteselim, or headman, who seemed very hospitably inclined, supplied me with food, of which I was much in need after my long fast of twenty-four hours, and endeavoured to prevail upon me to remain in his house until the weather had improved. He assured me that in consequence of the heavy fall of snow I could not cross the mountain; but, finding that I persisted in my resolution to proceed, he ordered a horseman to accompany me.

I was compelled, in consequence of the snow and hail, to take refuge more than once in huts which we found on our way. The snow was everywhere so deep that my guide had much difficulty in following the track over the mountain. After struggling for some hours up the steep ascent, we came late in the afternoon to a small khan. As there was no other shelter to be found before reaching the summit of the pass, we had to spend the night there. It contained two small rooms, which were crowded with travellers, who, like myself, were going to Beyrout, and had been unable to cross Mount Lebanon. They had stabled their horses and mules in these rooms, which were consequently warm, but close and filthy. As there was no one in charge of the building, it was impossible to obtain anything to eat there, and I had to be content with a little dry bread which I had put into my saddle-bags, and some 'dibbs,' or molasses made of grapes, which I obtained from a muleteer.

Sleep was impossible in the crowded room, in which I was scarcely able to lie down, and in which the muleteers discussed during the whole night the state of the weather and the possibility of reaching Beyrout. When morning came I found that the building was almost buried in snow, which was still falling. The horseman who had accompanied me from Zahlé declared that it was impossible to reach the top of the pass, and that it would be dangerous to attempt to do so. His opinion was shared by the other travellers in the khan, and he refused to proceed. I was, however, determined not to be baffled, and leading my mare by the halter made my way for some distance alone, when, losing all traces of the path, I could proceed no further, and was forced to return.

At the khan, a man who was from one of the villages on the mountain offered for a few piastres to show me the way, and as one of the travellers was willing to accompany me I resolved to make another effort. After struggling for three hours through the snow, and having to drag our horses out of the drifts into which they constantly fell, we reached the summit of the pass. The road on the other side had been in many places completely carried away by the rain, and we had the greatest difficulty in leading our weary animals down the rocky descent. As we reached the large village of Hammein, halfway down the mountain, the snow and hail suddenly ceased. A soft westerly breeze blew aside the clouds in which we had hitherto been shrouded, and disclosed a glorious expanse of blue sea, the far-stretching gardens of Beyrout and the town itself in the distance beyond. On a green plateau near the village a number of young men, gaily dressed and mounted on handsome horses richly caparisoned, were playing the 'jerid.'

After stopping for a short time at Hammein to rest my mare, I continued the descent of the mountain by a precipitous path in the direction of Beyrout. My fellow-traveller had left me, but I was now on a beaten track, and had no difficulty in finding my way. The sun shone brightly, and after the cold that I had experienced since leaving Damascus, the sudden and complete change of climate—from winter to summer—was as striking as it was grateful to me. It was dark before I found myself in the gardens of Beyrout, and eight o'clock before I succeeded in finding my way through its narrow, deserted, and unlighted streets, to the small inn in which I had lodged when passing through the town some months before.

I had already lost so much time that I could spare no more if I were to join Mr. Mitford at Aleppo. I accordingly started again on the following day. I passed an hour at the mouth of the Nahr-el-Kelb to examine the remarkable Assyrian sculptures and inscriptions carved on the face of the rock there.[3] Riding along the shores of the

[3] These sculptures were then supposed to be Phœnician. After the discoveries at Nineveh, the inscriptions, which are in the Assyrian cuneiform character, were deciphered, and found to relate to the wars of Sennacherib, who is represented in a tablet carved

beautiful bay of Jouni, I remained for the night at the small village of that name.

The direct road to Tripoli, which was carried along the coast, and close to the sea, was rocky and very trying to my mare. I made slow progress, and could get no further than a wretched khan situated in a marsh, and filled with Egyptian soldiers. The place was too foul for me to sleep in. I bought a few dried figs and some cakes of stale unleavened bread from the khanji,[4] spread my small carpet outside the building, tethered my horse close to me for fear of thieves, and passed the night in the open air.

I reached Tripoli on the following morning, and remained there for the remainder of the day to rest my mare. My road to Aleppo lay through Homs. I found no difficulty in travelling alone, notwithstanding the warnings which I received from the British Vice-Consul, a native of Syria, who seemed to think of little else but robberies and outrages alleged to have been committed by insurgent Mutualis to within a short distance of the town. He had made up his mind that I should be murdered on the road, and solemnly

on the rock. In the same place there are the remains of Egyptian, Greek, and Roman inscriptions.

[4] The words 'khan' and 'caravanserai' (properly karwan-serai), are used indiscriminately for a place of rest for travellers. The 'khanji' is the man who keeps the khan.

begged me to relieve him of all responsibility for the fate which inevitably awaited me. We were now in the midst of a Syrian spring. The face of the country was covered with the richest verdure, and enamelled with countless flowers. The air was soft and balmy; the sky intensely blue. The only drawback to the exquisite pleasure which I enjoyed in finding myself wandering alone through this beautiful scenery, without the impediments of servant and baggage, or the hindrance of an escort, was the state of the country, which had been reduced to a vast swamp by the recent heavy rains. I had to wade through it, with the water frequently reaching above the girths of my saddle, and my poor mare having to flounder and struggle through the mud. I occasionally met a solitary traveller or a small caravan. But no Mutualis nor other robbers were to be seen, although reports of their depredations and misdeeds were rife enough.

It took me three days to reach Hamah. The first night I passed in the ruins of an old castle in the hills, inhabited by a few poor Mutuali families. I have not kept a note of its name. The second I spent among the picturesque remains of the fine mediæval castle of El Hosn, surrounded by wooded slopes. A small village had been built near it by Mutualis, who treated me very hospitably. In the middle of the day I had rested at a large convent belonging to the Maronites. I found its long vaulted entrance filled with monks, who were anticipating an attack from the Mutualis, and were anxious to learn whether I had fallen in with any armed bands of them on my way. They appeared to be somewhat reassured when they learnt that I had travelled alone through the country, and had met neither with marauders nor with solitary robbers or thieves. The Superior regaled me with excellent wine of his own making, and wished me 'God speed' on my journey— very doubtful whether I should ever get to the end of it. The only name that I could obtain for the place was El Der, or the Convent.

I was under the necessity of remaining for a day at Hamah, as my mare had cast a shoe and was disabled in consequence, besides being much fatigued. I lodged there with a Signor Biazzi, an Italian—a military instructor in Ibrahim Pasha's army. My mare being still somewhat lame, I could make but slow progress after leaving Hamah, and it was only on the fourth day that I reached Aleppo, the country being in many places a swamp, caused by the winter rains, through which I had frequently no little difficulty in making my way. I passed the remains of many ancient buildings, tombs, and Christian churches; but I had no time to examine them, nor could I, under the circumstances in which I found myself, have made any notes of them, nor, as I was quite alone and without a guide, could I ascertain their names. The rolling plains which I crossed were very thinly inhabited. I had to sleep, or try to sleep, in miserable hovels filled with vermin. I could get little to eat, and little provender for my horse, which I had to tend myself. But in high spirits and my own master, I thought nothing of the privations and fatigues I had to endure.

At Aleppo I found Mr. Mitford, who, tired of waiting for me and not knowing what had become of me, was making his preparations to leave. He consented to remain there for a few days more, to give me and my mare a little rest, of which we were both very much in need. On March 18 we left Aleppo together, and reached Baghdad on May 2. I have related in my 'Nineveh and its Remains'[5] how, in descending the Tigris from Mosul on a raft, I visited the ruins of Nimroud, and first formed the design of making excavations in them should I ever have the opportunity of doing so. As my companion has described our journey in his 'Land March from England to Ceylon,' to which I have referred, I will resume my narrative from the time when we left that city.

[5] Chapter i.

CHAPTER VI.

Leave Baghdad—Join a caravan—Incidents on our march—Village of Yakubiyeh—Kizilrobat—Khanikin—Our travelling companions—Ruins of Holwan—Meet a French Ambassador—Sculptures at Ser Puli Zohab—The Ali-Ilahis—Cross the Persian frontier—Kirrind—The Lurs—Reach Kermanshah—Sculptures of Taki Bostan—Persian fanaticism—Difficulties at Kermanshah—The Governor—A Mûnshi—Continue our journey—The sculptures of Bisutun—The Shah's camp—The Hakim-Bashi—The Minister for Foreign Affairs—The camp raised—The Shah—Reach Hamadan—French officers—The Prime Minister—Hussein Khan—Difficulty in obtaining firman—The Baron de Bode—Cuneiform inscriptions—Separate from Mr. Mitford.

THE time had come for us to leave our kind friends who had made our residence at Baghdad so pleasant and so interesting, and to enter upon the most difficult and dangerous part of our long journey—that through Persia and the little-known regions between that country and Hindostan. I had determined to assume the Persian dress. Although I had been industriously studying the Persian language during my residence of nearly two months at Baghdad, my acquaintance with it was not, of course, sufficient to enable me to disguise my

European character. But I was advised that by wearing the native costume I should attract less notice, and consequently be less exposed to danger, and be less liable to insult and annoyance from the fanatical populations through which I should now have to pass. The Turks, Sunnis in religion, were less hostile to Christians and Europeans than the bigoted Shi'a Persians. In most parts of Turkey the people were accustomed to the European dress, as it had been to a certain extent adopted, with the addition of the scarlet fez, by the Turkish nizam or regular troops, and by the Ottoman officials. But in Persia European travellers had been rarely seen—in many parts not at all; and to have travelled in a costume which Shi'as considered indecent and almost insulting to their faith, might have exposed me to serious trouble, if not to danger, in remote parts of the country. I accordingly threw aside my Turkish dress which I had hitherto worn in travelling—it was well-nigh in rags from long and rough use—and replaced it by the long flowing robes, confined at the waist by a shawl, shalwars or loose trousers, and the tall, black lambskin cap, or 'kulâh,' then universally worn by the Persians. In addition I wore when riding a pair of baggy trousers of cloth, tied at the ankles, into which the ends of the long outer garment were thrust. My scanty linen, stockings, and shoes were also after the Persian fashion, and later on I shaved the crown of my head, leaving a ringlet on each side, and dyed my hair and beard a deep shining black with henna and 'rang.' I could thus pass very well, so long as my mouth was closed, for an orthodox Persian.

We had no servant with us, but trusted to the chance of finding some one on the way who, for a small remuneration, would give us such little help as we might require.

The country through which we had to travel was considered unsafe, as it was exposed to forays from the wild robber tribes inhabiting the mountains of Luristan, forming the boundary of Turkey, which we had to cross. We, therefore, thought it prudent to join one of the caravans constantly passing between Baghdad and the interior of Persia. Such a caravan was about to leave the city for Kermanshah—a large Persian town not far distant from the frontier. We accordingly made a bargain with a muleteer to supply us with two mules as far as that place. As our baggage was now reduced to the smallest possible compass, we did not require an additional animal for it. It was contained in our saddle-bags, which we could place across the high pack-saddles upon which we rode. Our carpets and quilts, thrown over them, made a soft and comfortable seat. They were somewhat difficult to mount, as we had only a loop of rope for a stirrup to help us, and we not unfrequently dragged them over and came to the ground.

As it was nearly the end of June, and consequently nearly the hottest period of the year, it was necessary to travel as much as possible by night and to rest during the day, to avoid exposure to the burning rays of the sun and the intense heat of the plains. The muleteers, with their mules and loads, and the travellers who formed the caravan, had been encamped for several days outside the walls of the city ready to begin their march. But there were the usual Eastern delays, and it is not easy to set a caravan in motion. At last, the 29th of June having been pronounced by a Mulla, after consulting the Koran, a propitious day for commencing a journey, our muleteer came to us in the afternoon with his animals, and announced that we should certainly start that evening immediately after sunset. We mounted our mules and, passing through the long and intricate bazars, left the city by the northern gate. We found our fellow-travellers on the banks of the Tigris, preparing for their departure. But at the last moment, when the animals were loaded and every one was ready, the Karwan-Bashi, or head of the caravan, refused, for some reason or another, to leave on that day. A vast amount of wrangling, yelling, and quarrelling took place, but he persisted in his determination, and would neither yield to persuasion nor to threats. At length a part of the company went on their way, leaving the obstinate muleteer behind. He promised to join them before they had reached the Persian frontier. As he was the owner of our mules we were compelled to remain with him, and as, before the dispute had come to an end, the sun had set and the city gates were closed for the night, we had nothing to do but to spread our carpets on the ground and to make ourselves as comfortable as we could.

We passed the following day with our friends in the city, and again joined the caravan late in the afternoon. The Karwan-Bashi was now ready to proceed. The mules were loaded and the travellers mounted. As soon as Mitford and I joined them we all moved off together as the sun went down.

The caravan, or 'khâfileh,' was made up of a motley company of Persians—petty traders with their wares; pilgrims with their wives and children, on their return from the holy cities of Kerbela and Kausimain—'Kerbelayis' as they are called in Persia; several 'hajis,' who had performed the pilgrimage to Mecca, and who consequently, like the 'Kerbelayis,' enjoy a kind of sacred character; and a few ordinary travellers on their way home. Some rode on horses, mules, or donkeys, generally perched on the top of their baggage; others went on foot. Most of the men were armed. The women were enveloped in 'chaders,' or ample mantles of silk or cotton—some richly embroidered with gold or silver thread—which envelop the whole person. Their features were concealed by the horse-hair veil generally worn by the ladies of Baghdad and of Persia.

No order was maintained during our march. There was no regular road, but we followed a broad track across the open country. As it was soon dark and the animals were mostly without bridles, having only a halter to guide them, we jostled one against the other, sometimes knocking over the loads and sending those mounted upon them sprawling on the ground. There were consequently constant cries and screams of women and children, and frequent stoppages to remount those who had fallen or to readjust the baggage. Some of the men smoked kaleóns, or water-pipes, which they lighted from iron pots filled with burning charcoal hanging to their saddles. Others sang Persian couplets in a loud, shrill, vibrating voice. Others again repeated verses from the Koran in a monotonous drawl, interrupted every now and then by a cry of 'Ya Allah! Ya Ali!' in which all the company joined.

We occasionally passed caravans going in the opposite direction to Baghdad and Kerbela. Those destined for the latter place were usually conveying dead bodies, enclosed in rude wooden coffins, swung in couples on the backs of mules. They were the remains of pious Shi'as sent by their relatives for burial near the holy shrines of the Imaums. We had no difficulty in recognising the nature of the loads, from the stench which they emitted.

As our animals were fresh and we were passing over a perfectly level plain, we proceeded at a smart walk. When we had marched about two hours a halt was called, and all dismounted and spread their carpets for evening prayer. The necessary ablutions having been performed, one of the party, a seyyid, or descendant of the Prophet, and a haji, acted as 'Pish-nemaz,' or prayer-leader. Standing in front he repeated the prayers in a loud voice, and went through the customary genuflexions and bowings, in which he was followed by the rest of the company. After this detention we resumed our journey. Occasionally we passed near villages which could just be distinguished by the palm trees surrounding them, showing darkly against the sky. The night was delightful, the stars shone brightly, and a refreshing breeze cooled the air after the intolerable heat of the day. The sun rose in unclouded splendour. The heat soon became so oppressive that we rejoiced when we reached, after two hours' more ride, the village of Yakubiyeh. We all dismounted, and little groups were formed under the palms, or in the shade of the mud walls that enclosed the gardens. Mitford and I found a pleasant shelter under some orange trees. Preparations were made for passing the day. Fires were lighted and breakfasts cooked. The villagers brought thin cakes of unleavened bread, 'ghee' or butter, sour milk, and dates and other fruit for sale. We had provided ourselves with a bag of rice and a couple of saucepans, and, like the rest, cooked our own food. After we had eaten, we followed the example of our fellow-travellers, and lay down to rest. The whole caravan was speedily buried in sleep.

The country through which we had passed during the night formed part of the vast alluvial plains which stretch from the Persian mountains to the river Euphrates, and, beyond, to the great Syrian desert. At this time of the year they were parched and devoid of vegetation—a dry arid waste, with an occasional Arab village surrounded by palm trees and by gardens cultivated by means of watercourses derived from the river

Diyala which, after many windings, joins the Tigris below Baghdad. There was nothing of interest to deserve our attention, except the remains of innumerable ancient canals, whose high banks and waterless beds constantly crossed our track.

At sunset we were summoned by the Karwan-Bashi, who directed all our movements and appointed our stations, to load our animals and to mount. In a short time we were again on our way, having slept for several hours and being prepared for another night's ride. We rode over the same flat, parched, and uninteresting country, and soon after sunrise sought shade and repose for the day in the gardens of the village of Nushirwan.

During the following night we crossed the Hamrin, a range of low crumbling sandstone hills which extend, parallel to the great mountains of Kurdistan, almost from the river Zab near Mosul to far below Baghdad. We halted for the day at the large village of Kizilrobat, in a vast caravan-serai, where we were able to procure provisions and to obtain shelter from the sun. We had made but a short journey of about five farsaks,[1] or fifteen miles, and did not leave our resting-place

---

[1] The farsak, the ancient Persian 'parasang,' is about three miles in length, or the distance that a horse or mule can walk in an hour.

until three hours after sunset, as we had only the same distance to perform during the night.

We travelled over rough stony ground, broken by hills, or rather hillocks, with very precipitous sides, over which we had to climb, the horses and mules frequently falling and throwing their riders, especially the women, who had to sit cross-legged almost without support on the mountain of carpets and coverlets piled upon the packsaddles. As the day broke we found ourselves approaching the foot of the great Zagros range, and soon after sunrise reached another large caravanserai at the picturesque but half-ruined village of Khanikin, embedded in palms and watered by the small river Holwan, here crossed by a large and well-built bridge. Behind the village rose abruptly the lofty mountains which we were about to enter. We spread our carpets in a delightful garden, amidst apricot, fig, pomegranate and other trees, which afforded a most pleasant and grateful shade, and on the bank of one of the numerous watercourses which brought water from the river for purposes of irrigation. We preferred this resting-place to a room in the dirty and crowded khan, as the owner of the garden, after a great deal of quarrelling, was satisfied with a small piece of money, and allowed us to supply ourselves abundantly with fruit.

During the following night we joined that part of the caravan which had left Baghdad the day before us. It was chiefly composed of Armenians, with their wives and children, who were on their way to Julfa, a suburb of Isfahan inhabited by that people. They were waiting for us, as there were rumours that there were bands of robbers from the mountain tribes on our road. Some travellers had been plundered a week before. The muleteers took alarm, and wanted to induce the people of the caravan to subscribe for a guard of armed men, which they said could be procured in the village, and which, they maintained, was absolutely necessary for our safety. However, no one would contribute. With the addition of those who had now joined us it was thought that we could muster a sufficient number of armed men for our defence, in the event of our being attacked.

During our long rides at night and our hours of rest in the day we had become acquainted with some of our fellow-travellers, who were helpful and obliging. In talking with them I made good progress in the Persian language, which they were always ready to teach me. They were for the most part traders returning from Baghdad with European goods for sale in Persia. The seyyids, the hajis, and the Kerbelayis kept aloof from us, for as Christians we were unclean and not fit company for such holy persons. The Persian Shi'as are much more fanatical in this respect than the Turks, who are Sunnis. They will not allow a Christian to eat out of the same dish with them, nor to drink out of the same vessel, nor to smoke the same pipe. The more bigoted will not even permit a Christian to touch their clothes or their hands, and will perform an ablution immediately after such contact. We could not borrow a drinking-cup from our Persian companions, nor a cooking-pot, and were compelled to rely entirely upon our own resources for such articles. When the Armenians joined us we were better off, as they had no such prejudices, but, on the contrary, were anxious to be of service to us, as they considered that, as Englishmen, we might afford them, in case of need, some protection. They were consequently ready to help us in cooking our food and in loading our mules.

We again made a short journey through a rocky and hilly country to Kasri-Shirin, a village in ruins and deserted on account of the depredations of the robber tribes dwelling on the Persian frontier, which were perpetually making forays in the plain, pillaging the villages and plundering caravans. During the night, notwithstanding the fears of our muleteers and an occasional panic owing to an alarm that horsemen were heard or seen in the distance, we met with no adventure. Had we been attacked by even a few mountaineers, our boasting Persian companions, who were constantly examining and firing off their matchlocks, and talking very big of what they would do in the event of any Lurs being sufficiently bold to venture to make their appearance, would, I was convinced, surrender at once, and allow themselves to be stripped to the skin without making the least resistance. The Persians are notorious cowards and boasters, and it frequently happened that large caravans comprising several hundred people were stopped and robbed by a few—perhaps not more than half a dozen—bold marauders from the mountain tribes. I was not sorry, therefore, that we crossed this dangerous borderland without putting our companions' valour to the proof.

We stopped for the day in a large caravanserai. After taking a delicious bath in the cool waters of the river Holwan, I explored the ruins of an ancient city, apparently of the Sassanian period, on a height within a short distance of the village. Its walls, constructed of roughly hewn blocks of red sandstone, united by cement, were well preserved. The houses, of which I found numerous remains, were built of rounded stones or boulders, from the river. Amongst them were the ruins of what appeared to have been a palace, or a khan.

The building consisted of a quadrangle of considerable size, formed by massive walls and having only one large entrance. I observed numerous underground chambers or vaults, similarly constructed of rounded stones, such as are very common in ruins of the period of the Sassanian dynasty of Persia. They may have served for dwelling-places, or may have formed the foundations or basements of buildings. The ruined city was called Shirin-i-Khusrau, as, according to tradition, it was built by Chosroes for the beautiful Shirin, the heroine of Persian love-stories. Near it were the remains of what appeared to be an ancient fort, probably intended as a defence to the pass into which the road now entered. It was erected near a ravine formed by the river Holwan, which forces its way through the hills, leaving lofty cliffs curiously stratified with alternate beds of white gypsum and red sandstone.

In the middle of the following night, when winding amongst the mountains, we were alarmed by the report that a large body of horsemen were approaching us. The muleteers and our timid fellow-travellers were persuaded that we were about to be attacked and plundered by marauders. They were much too frightened to think of resisting, had their fears proved well founded. However, we soon discovered that the party descending the mountains consisted of the French Ambassador, M. de Sercey, and his suite, with a large escort of irregular cavalry, returning to Baghdad from an unsuccessful mission to the Shah. We had been led to expect that we should meet him on our road, and had been charged by some members of his embassy, who had preceded him, with a packet of letters for him, which I accordingly delivered. After a few minutes' conversation he continued his journey and I rejoined the caravan.

About two hours before daylight we reached the village of Ser-puli-Zohab. The caravan did not halt at this place, but as there were ruins and sculptures in the neighbourhood which we were desirous of examining, we prevailed upon our muleteer to stop there for some hours. The ruins consisted principally of numerous mounds, marking the site of an ancient city. But there were no remains above ground, and the only object of interest which we could discover was a rock-cut tomb on the face of a precipice, which has been scarped to the height of about eighty feet. Beneath the entrance to this tomb, carved on the rock in low relief, was a figure representing a 'mobid,' or high-priest of the Zoroastrians, with one hand raised as if in the act of benediction, and holding in the other what appeared to be a scroll, probably intended for the sacred leaves of the Zend-Avesta. He was dressed in the pontifical robes worn by the Zoroastrian priests, with the square cap pointed in front, and a kind of hood covering the ears and lower part of the face.[2]

To enter the tomb a long ladder was required, and none could be procured. As far as I could see from below, it appeared to resemble other rock-cut tombs of the same period, so frequently met with in the mountainous parts of Western Persia. The entrance had been ornamented with two columns carved out of the rock. It may have been the place of sepulture of some local prince or chief during the period of the Kaianian dynasty. The present inhabitants of the country know it as the Dukkani-Daûd, or David's Workshop—that Jewish monarch having, according to their traditions, been a blacksmith—and regard it with great veneration. Our guide, when approaching the rock, frequently prostrated himself, kissed the ground, and called upon the Prophet David for protection.

The inhabitants of the district of Zohab, which we were traversing, are of a very marked type, distinct from that of the tribes by which they are surrounded. They are believed to be the de-

---

[2] For a full account of the ruins and sculptures of Holwan and of the Ali-Ilahis, the reader is referred to the very interesting paper by Sir Henry Rawlinson in vol. ix. of the *Journal of the Royal Geographical Society.*

scendants of some ancient race which has occupied these mountains from a very early period, or which had been driven to seek refuge in them from the plains by foreign invaders. As they preserve numerous traditions with which the name of David is connected, it has been conjectured that they may be a remnant of the Lost Tribes, and they have appeared among the numerous claimants to that distinction. In religion they are Ali-Ilahis, professing to be Mohammedans, but held by true believers to be infidels of the very worst description. As their name denotes, they are said to worship the Imaum Ali, the cousin of Mohammed, whom they believe to be one of the thousand-and-one incarnations of the Deity. Similar sects are found in other parts of Western Asia. The Ali-Ilahis of Zohab are also known by the name of 'Chiragh Sonderan,'—the putters-out of lights—from certain mysterious and impure rites which they are alleged to practise in the dark. But similar calumnies have been spread against other religious sects by ignorant fanatics, not only in the East but in the West, to justify persecution. As is well known, the early Christians were not exempt from them. The poor Ali-Ilahis are probably equally innocent of the charge. However, they bear a very bad reputation on the score of morality, and according to general report lead very dissolute lives. The dancing boys and girls who frequent Baghdad, and are notoriously of evil fame, come principally from this district. Whilst we were resting at the caravanserai a party of them came to perform their indecent dances before us, as they were in the habit of doing on the arrival of travellers.

The river Holwan issues at Ser-puli-Zohab from a deep gorge through lofty precipices. Cut in the rocks forming its entrance are several tablets on which figures have been carved in relief. They are much defaced by exposure to the weather, and I could distinguish but two of them. There are others which can only be seen at certain times of the day when the light is favourable, as they are almost obliterated. On that which I examined there had been six figures, but only one remained —that of a warrior with a shield and club—and there were traces of a Greek inscription, with the word ΒΑΣΙΛΕΥΣ and two or three letters. On a lower tablet were rudely represented a man on horseback and another on foot, with fragments of two inscriptions in the Pehlevi, or Persian character of the time of the Sassanian kings.

The ruins at Ser-puli-Zohab are those of Holwan, a very ancient city, which after the Mohammedan occupation still contained a considerable population.

After examining these monuments we passed through the gorge formed by the river and entered a small plain, in which were several encampments of Ilyats,[3] with their black tents and their flocks. They were Lurs, in search of pasture. We found that our caravan had stopped near a solitary tree. As this tree afforded the only shelter, and the women and children had been gathered under its shade, we had to pass the rest of the day exposed to the burning rays of the sun. We resumed our march early in the afternoon, and rode for about twenty-five miles through a mountainous and difficult country to the large village of Kirrind. We stopped on the way for half an hour to rest the animals, at a caravanserai built at the foot of a very steep ascent. The road was carried along the face of the rock on the edge of a precipice overhanging deep ravines. It was a mere mule track, perilous even by daylight. At night, with the animals left almost to themselves and hustling one another, our position was far from pleasant, and we ran some danger of being precipitated into the abyss below. This was, however, the high road between the plains of Babylonia and the highlands of Media, upon which, centuries ago, there must have been a vast traffic, and which is still frequented by caravans engaged in the trade between

[3] The Persian name for nomad tribes living in tents.

the provinces of Southern Turkey and Northern Persia, and by the numerous pilgrims who constantly visit the shrines of the Imaums at Kerbela and other holy places in Mesopotamia. At one time a properly constructed causeway probably existed, by which that traffic was carried on by carts as well as by beasts of burden. At about two-thirds of the ascent there is a small square building of large dressed blocks of white marble, consisting of a deep vaulted recess, which is Greek or Roman. It has been called a gate, which it is not. It may have been a station for the collection of tolls or black-mail upon men and merchandise, or a temple in which persons travelling over the mountain offered up sacrifices for their safety to the gods. Zagros has at all times been inhabited by warlike robber tribes—a terror to travellers. The natives call the edifice the Taki-Girrah—the arch holding the road. The pass itself is now known as the Gardanai-Taki-Girrah, or as that of Kirrind, from the village at its eastern foot, but by the Arab geographers it is named Akabahi-Holwan, or the defile of Holwan. On the summit is a large caravanserai and a village named Surreh-Dereh. Here we crossed the Persian frontier, and then descended rapidly to the very pretty village of Kirrind, situated in a deep gorge, formed by mountains rising precipitously on every side. We approached it by an avenue of tall poplar trees, through vineyards and gardens filled with fruit trees of various kinds. A copious stream bursting from the rock and divided into innumerable rivulets irrigated the cultivated land and clothed the valley with verdure. The caravanserai was a large and handsome quadrangular building, consisting of a spacious courtyard surrounded by stables for horses and mules and small rooms for travellers, which were dirty and abounding in vermin. For their occupation a trifling fee was paid to the khanji. This was the first Persian village I had seen. I was charmed by its pretty and cheerful appearance, surrounded as it was with trees and gardens.

Our night's march was the longest and the most fatiguing that we had performed since leaving Baghdad. We had been above fourteen hours on our pack-saddles, and I was not sorry when the head of the caravan announced that we were to remain the following day at Kirrind to rest our wearied animals, which had suffered not a little from their toilsome journey over the Zagros pass. The air was cool and refreshing, and offered a delightful contrast to the sultry heat and burning winds of the Mesopotamian plains. I spread my carpet in the shade of a tree in one of the gardens.

We left Kirrind soon after midnight on the second day, and reached the village and khan of Harounabad an hour after sunrise. There were several encampments of Ilyats hard by. They brought us fresh unleavened bread baked in thin crisp cakes, 'âb-doogh,' or sour milk, and fresh butter. These luxuries were very enjoyable after our scanty fare during the previous days. My companion and I had to prepare our own dinner, which usually consisted of boiled rice and a little meat when we could obtain it, which was not often the case, from the keeper of the caravanserai. We collected such wood or dry grass as we could find, unless the inhabitants of a neighbouring village brought fuel for sale, which we purchased for one or two copper pieces. We drank nothing but water, except when we could procure sour milk, the common beverage of the nomad tribes. An excellent appetite, after a long night journey, made us relish our simple meal, and I was never in better health, notwithstanding the fatigue we underwent—and I know nothing so fatiguing as travelling by night over rough and stony roads on a jaded horse requiring constant urging.

We were now traversing a part of the province of Luristan chiefly inhabited by semi-independent Lur tribes. They are of the ancient Persian stock, and for the most part nomads, living in the mountains during the summer and descending into the plains with their flocks and herds in the winter. They have an evil repute as highway robbers and arrant freebooters. They are in constant rebellion against the Shah, and cause the Persian Government no little trouble by plundering caravans and making forays on the territories of their neighbours, the Turks. At the time of our journey, two of the tribes being at war, the country was considered very unsafe for travellers.

The Lurs wear the Persian costume,[4] with the

[4] At that time the dress of a Persian consisted of the following parts: The *Kuláh*, a tall conical cap, or hat, of black lambskin, with the upper part turned in—those from Bokhara with small thick curls were the most prized. The *Khabu*, a long robe of silk or other material reaching to the ankles and open in front, with the skirts divided in two downwards from the waist, where it was confined by a shawl folded round the body. At the arms it was usually left open, and fell in lappets, which could be buttoned up if necessary, as they always were in the presence of a superior. Over the *Khabu*, a coat of cloth or of Cashmere shawl—of great value in the case of persons of rank—fitting close to the body, as far as the knees and the elbows, was generally worn. The *Arkoloch*, an under-dress, made like the *Khabu* but of light cotton-stuff, generally of many colours; the *Pirakan*, a short shirt of white linen, descending to the waist and buttoning across the breast; the *Shalwar*, or loose trousers, very wide at the ankles; and the *Sirjameh*, or drawers of linen, completed the attire. Stockings, white or worked in coloured wools, were worn; and shoes, or slippers, except in riding, when high loose boots were used. Since I was in Persia a black conical cap of cloth has been generally substituted for that of lambskin. I am not acquainted with the other alterations

exception of a close-fitting skull-cap made of white felt—a much more convenient head-dress for riding and for fighting than the tall lambskin hat then worn by the Persians. They always go armed with their long matchlocks, and with a pistol and hanjar, or dagger, in their girdle.

After leaving Harounabad we again plunged into mountains thinly wooded with dwarf oak, travelling, as usual, by night, and crossed a high and difficult pass by a very bad and rocky road. It bears the name, which it well deserves, of Gardanai-Nal-Shikan, or the Pass of the Broken Horseshoes. We spent the day in a deserted khan in a well-irrigated plain rich in herbage, called Maidasht, in which were numerous encampments of Lur Ilyats.

A short ride next day of four farsaks brought us to Kermanshah, the first town that we had seen in Persia. It stands in a fine, well-watered plain, surrounded by lofty serrated mountains towering one above the other, with high and precipitous peaks, then still covered with snow. It is a place of considerable size, in the midst of gardens, vineyards, and orchards, amongst which are wide-spreading walnut trees and lofty

---

made in the national dress, which, I believe, has been to some extent changed or modified, especially in the cities, where European influence has prevailed in this and other respects.

poplars. An abundant supply of water descending from the mountains, divided into numerous canals, irrigated the lands, and rendered them bright with verdure. Altogether I was very favourably impressed with the appearance of the place from a distance. I thought it one of the prettiest and most flourishing towns I had seen in the East. But on entering the gates I found that a great part of it was in ruins, although the bazars were extensive and well supplied with native produce and foreign goods. We hired a room in a khan, called that of the Aga, which was not over clean or free from vermin, and spread our carpets in it. We got an excellent dinner at a very small cost from an eating-house. It was brought to us on a large circular pewter tray, and comprised soup, one or two made dishes with savoury sauces, a very good pilau, delicious fruit of various kinds, including very fine apricots and plums, and an iced and delicately flavoured sherbet. This was our first introduction to Persian cookery, which we thought far superior to that of Turkey.

At a short distance from Kermanshah are the celebrated sculptures of Taki-Bostan. We rode to them the day after our arrival, and found encamped near them M. Flandin, a painter, and M. Coste, an architect, both attached to the mission under the French Ambassador whom we had met on our road. They had remained behind to make drawings of these monuments. They received us very courteously, and we spent the day with them. The great work published by the French Government on the return to France of these gentlemen contains elaborate representations of the rock sculptures, which are of the time of two kings of the Sassanian dynasty, Shapur Dhu-l-aktaf, and Shapur, the brother of Bairam.[5] I made some sketches of them and copied the inscriptions in the ancient Persian or Pehlevi character. The bas-reliefs consist principally of scenes from the chase, carved most elaborately and with great spirit. The King and his Court are seen hunting stags, wild boars, and other animals, using in their pursuit elephants, horses, and camels. For a full description of these remarkable monuments I must refer the reader to Mr. Mitford's book and to M. Flandin's drawings.

During our journey from Baghdad we had suffered constant annoyance, and even insult, from some of our fellow-travellers, and especially from the mullas, seyyids, and other fanatical fellows, who were returning from the pilgrimage to Mecca or Kerbela, and who considered that they had acquired a character for sanctity which forbade their mixing with such infidels as ourselves. They avoided

[5] See Flandin et Coste, *Voyage en Perse*, Paris, 1851.

us as much as possible, and invited others to do the same. They would not allow us to spread our carpets near the spot which they had chosen for their day's rest. They jealously watched the water jars which were brought to us by the villagers, when we arrived at a camping place tired and thirsty, lest we should drink out of them or touch them. The women were made to pull down their veils whenever we approached them, and even the children were taught to run away from us as if we were infected with the plague. The chief of our persecutors was a mulla from Meshed, who had a great reputation for holiness among his companions. As we rode along in the night he was constantly uttering pious ejaculations and curses directed against us, intermingling them with verses from the Koran. But he was even more violent against a Musulman from Cabul, a Sunni, who was friendly to us during the march and helped us in many little ways, and whom he denounced and cursed in no measured terms for associating with 'Kâfirs' or infidels, and becoming thus himself unclean. This was considered a crime of the deepest dye by orthodox Shi'as. The Afghan had been 'mûnshi'—a kind of secretary for the Persian language—to Colonel Hughes, the governor of the island of Karak, then occupied by England, but had, I afterwards learnt, been dismissed on suspicion of communicating to the Persian authorities information which he had acquired in his official capacity. He had joined our caravan at Baghdad, and was on his way back to Afghanistan. He had undertaken to give me lessons in Persian during our journey, on condition that I paid his expenses on the road, which amounted to little enough. But he was soon frightened by the denunciations of the mulla, and would neither eat nor drink with us out of the same vessel, and became somewhat shy at being seen in our company by the hajis and Kerbelayis.

He was, however, of use to us in our quarrels with our fellow-travellers. They had become insolent through the provocation of the mulla, and frequently drove me into threatening acts of violence with the butt end of my gun, or with a stout stick with which I had armed myself for the purpose. He would interpose or mediate before we came to blows, and thus prevented quarrels which might have led to bloodshed and even more serious results. A man would kick away my carpet because in his opinion it was spread too near his own. On one occasion I was struck by a fanatic whose religious feelings were offended by something that I had unwittingly done. Had it not been for the intervention of the mûnshi I should have broken his head.

It was evident that as soon as we had crossed the Persian frontier we should be exposed to many troubles and difficulties. The Persians were then less accustomed to seeing Europeans than their neighbours the Turks, and were taught to look upon them, as well as on all Christians, as unclean, and as fit objects for insult and aversion. They were wont to speak of them as dogs and pigs, animals which good Shi'a Musulmans carefully avoid, and by whose touch—were it even only to a part of their garments—they would be polluted and would require ablution. We found ourselves subjected to a kind of quarantine, and there was a very bad feeling springing up against us.

Moreover, at this time, although England was not actually at war with Persia, she had suspended her diplomatic relations with the Court of Tehran, had withdrawn her ambassador, Sir John MacNeill, and had occupied the island of Karak in the Persian Gulf, belonging to Persia. It was expected that actual hostilities would soon break out between the two nations, and the air was full of rumours of war. We were, therefore, running some risk in venturing into what might be considered an enemy's country, and it was to be expected that our travelling companions would denounce us, as they knew that we were Englishmen, to the Persian authorities.

I had received letters from Suleiman Aga, a Persian gentleman at Baghdad, for one Mohammed Jaffer Aga and his son Abd'ullah Aga, who were notables of Kermanshah. I was also charged with a letter for them by Suleiman Mirza, one of the refugee Persian princes whose acquaintance I had made at Baghdad. I called upon them the morning after our ride to Taki-Bostan. I was accompanied by the mûnshi, as interpreter. The Agas were mullas, and I found them seated in a handsome hall, surrounded by a large company of their brethren. They did not receive me very politely, and I soon took my leave. The mûnshi had imprudently delivered the letter from the prince, who had been banished from Persia, to Mohammed Jaffer Aga before the assembly, and this probably had caused him some annoyance.

In the evening, as Mitford and I were seated on our carpets in the little room we had hired in the khan, one of the secretaries of the governor of the town called upon us. He intimated that his master was surprised that we had not been to see him. We made the best excuses we could devise—that we were simple travellers, not in a condition to present ourselves to a man of his exalted rank, that we feared to disturb him, &c. However, we expressed our readiness to visit him, should he be disposed to receive us. It was agreed that the secretary should call upon us early on the following morning to conduct us to the governor. After he had left us we saw him in close conversation with our friend the mûnshi, whom he was evidently questioning about us. Next day, as early as six o'clock in the morning, the secretary, accompanied by a number of followers, came to take us to the governor. We walked with him through the bazars—our cortège increased as we went along by idlers who perceived that there was something unusual taking place—to the palace, the exterior of which, like that of most buildings, public and private, in Persia, was of plain mud-built walls, without windows or architectural decorations. However, after passing through a long, dark passage, with raised seats of brickwork on either side, upon which guards and attendants were lolling upon their carpets, smoking kaléons or water-pipes, we emerged into a spacious courtyard with a large 'haush' or reservoir of clear water in the centre, surrounded by roses and other flowers in full bloom. The building was of one story, and of the picturesque architecture of Persia. Halls, entirely open on one side to the air, had once been gorgeously and elaborately painted. Smaller chambers were closed by beautiful lattice-work. The whole was, however, in a ruined condition.

Here again were guards and servants idling about. Passing through a second narrow passage we found ourselves in an inner court, with a fountain and a parterre of roses in the centre. At the further end of this court, which was enclosed by low buildings like the first, in an open recess, and surrounded by vases filled with sweet-scented flowers, sat, squatting on his hams, the governor Serdar, or general, Nur-Mohammed Khan. He was a man of middle age, and of handsome appearance, with a long beard dyed the deepest black. He wore a coat of the rarest Cashmere shawl, and a high cap of the finest lambskin, and carried a jewel-hilted dagger in the shawl folded round his waist. By his side sat his son, a remarkably pretty boy about ten years of age, also very richly dressed.

The secretary, advancing to the recess in which was the Khan, made several profound bows, his long beard almost touching the ground, and announced our presence. We were directed to ascend to the place in which the governor was sitting. He received us coldly, and without any of that high-bred courtesy which we had almost universally experienced from Turkish officials and men of rank. Smoking his kaléon, and occasionally sniffing at a raw cucumber which he held in his hand, he put a series of questions to us, which were translated by the mûnshi, who continued to stand below. Having ascertained that we were English travellers, whose only object was to pass through Persia on their way further east, he asked us whether we were furnished with any special permission to enter the country at that time. We produced our passports, which we had taken the precaution to have translated into Persian before we left Baghdad. He appeared to be satisfied as to our respectability, and as to our character of *bonâ fide* travellers, but he informed us that considering the suspension of relations between Persia and England, and the withdrawal of the English Ambassador from Tehran, we could not be permitted to travel in the country without a special permission from the Shah, and that we must remain at Kermanshah until it could be obtained. We then suggested that we might be permitted to go ourselves to his Majesty, who was encamped at only three days' distance, under the surveillance of any guard that he might think proper to send with us. To this proposal he acceded, and we left him with the understanding that we were to leave on the following morning.

However, we had scarcely reached our khan when a soldier arrived with instructions from the Governor to keep watch over us, and to prevent us from leaving the town. He had altered his mind, and would not allow us to proceed on our journey until he had been authorised by the Shah to do so. We were informed that an answer would be obtained from his Majesty, to whom a special courier had been sent, within two days. There was nothing to be done but to submit. Not to waste our time I determined to revisit the sculptures of Taki-Bostan, but we were unable to procure either horses or mules.

Several circumstances led me to suspect that our companion, the mûnshi, whom we had trusted, had betrayed us, and that it was through his communications to the Persian authorities that our arrival at Kermanshah had been made known to them, and that we had been detained almost as prisoners. We discovered that he had been spreading reports that English troops were being assembled at Karak, with a view to the invasion of Persia and the dethronement of the Shah, who was to be replaced by one of the exiled princes then residing at Baghdad; and that to this end the English were sending secret agents into Persia with large sums of money to corrupt the Persian troops and to obtain partisans. There was every reason, therefore, to believe that we were so employed, and that we were travelling with these objects. The letter to Mohammed Jaffer Aga from Suleiman Mirza, which we had incautiously confided to the mûnshi to deliver, tended to confirm these suspicions.

We also found that he had prevented our procuring horses, by telling their owners that they would be bastinadoed by the governor if they ventured to hire them to us. His motive in having us thus detained was not quite clear. He might have wished to revenge himself upon us as Englishmen for his dismissal from the service of Colonel Hughes; or, what seemed to be more likely, he did not want us to continue our journey until he was ready to accompany us, as we had promised to pay his expenses, and he had brought a quantity of wares to be sold on his way home a part of which he hoped to dispose of at Kermanshah. These wares consisted of relics from the places of pilgrimage—soap from Mecca; circular bits of earth from Kerbela, upon which devout Shi'as place their foreheads when prostrate in prayer; rosaries from the holy cities, and other merchandise of that nature, which always meets with a ready sale in Persia at a very considerable profit.

As it was important that we should remove the suspicions which through the mûnshi's treacherous conduct attached to us, we determined to return at once to the governor to offer explanations. We could not trust the mûnshi to interpret for us, so we availed ourselves of the services of one Saleh, a Lur, who had been in the service of Dr. Ross, of Baghdad, and who spoke a little Arabic.

The mûnshi followed us to the palace, and, fearing that Saleh might prove a witness against him, endeavoured to prevent us having a hearing from the Serdar, protesting against our interpreter being allowed to appear in the presence of the governor, on the ground of his being an Armenian, and consequently an unclean infidel. But the Lur was very indignant at his religion being called in question, and after roundly abusing the mûnshi and repeating the Mohammedan profession of faith, swore by Allah that he was a true believer. No greater offence can be committed against a Musulman—an offence which all good Mohammedans are bound to resent—than to accuse him of being a giaour, or infidel. The mûnshi was consequently alarmed at the consequences of his calumnious charge, which would have brought upon him condign punishment from all true believers, and slunk away, leaving us with our new interpreter. Nur-Mohammed Khan, who had all the airs and graces of a 'petit maître,' was more amiable and condescending than he had shown himself on the occasion of our first visit. He received our explanations patiently, declared himself satisfied with them, and expressed his willingness to allow us to proceed at once under escort to the Shah's camp at Hamadan. As in consequence of the mûnshi's proceedings no one would give us horses for our journey, we asked that an officer should be sent with us to enable us to procure them. To this he also assented.

We were accompanied on our return to the khan by this officer and by several of the governor's attendants, who assembled in our room and seated themselves on our carpets, ostensibly to help us to procure horses, but really for the purpose of learning news and of gossiping. In Persia it was the custom for a person before entering a room to take off his boots or shoes, and to leave them at the entrance. Like the rest of the company, I had left my boots at the door. When the party broke up and dispersed they had disappeared, having evidently been stolen. The thief could only have been one of the Serdar's officers or secretaries, as no one else had been with us. I threatened to go at once to the palace to lodge a complaint.

The soldier who had been placed as a guard over us pretended to be very indignant, and swore by his beard that he would discover the thief and bring him to me. He went away in search of him, and soon returned with one of the governor's officers who had been sitting with us, a well-dressed and highly respectable-looking man, who declared that he had nothing whatever to do with the matter, and offered to swear on the Koran to his innocence. However, he produced a pair of boots similar to mine, but only much older and very much the worse for wear, and proposed that I should accept them in lieu of those I had lost. He was, no doubt, the thief, and I declared my determination to lodge a complaint against him, and proceeded to the governor, whom, however, I could not see, as he was taking his siesta. On my return to the khan I found that my own boots had been slipped into the room whilst I was away, and whilst Mitford was asleep. This was my first experience of Persian dishonesty, and I was taught by it to look sharp after my property.

Our guard had behaved well in this matter, and although he thought it necessary not to let us out of his sight and to accompany us wherever we went, he walked before us through the crowded streets, and cleared the road for us, dealing lusty blows with his stick upon man, woman, or child who did not quickly get out of our way. We amused ourselves during the time we were detained by visiting the shops in the bazar, especially those for the sale of objects of Persian manufacture, such as enamels, wood-carvings, embroideries, carpets, and other articles of this nature, in which I took much interest. They showed an elegance of design, a true feeling for colour, and a skill in execution which gave me a high opinion of the taste and fancy of Persian artists and workmen. In those days they had not been spoilt by contact with Europe, and by an attempt to imitate vulgar European manufactures in colour and designs.

One of our fellow-travellers from Baghdad, a certain Mirza Ismail, a native of Shiraz, had been an exception to those who had treated us with so much incivility and insolence. He was lodging during his stay at Kermanshah with a well-to-do shoemaker, who asked us to be his guests at supper. We accepted his invitation and had an excellent entertainment. However, although our companion and his host were intelligent and apparently liberal-minded men, they would not allow us to eat out of the same dish with them, and we were served apart, which we had no reason to regret, as there were neither knives nor forks, and every one helped himself from the same dish with his fingers. After supper we were served with finely-flavoured tea, with lemon-juice instead of milk, and with kaleóns filled with the finest Shiraz ' tumbaki.'

We had been promised horses for an early

hour in the morning, but they did not arrive until late in the afternoon. The horseman who was to keep watch over us until we reached the camp of the Shah and were handed over to the authorities there, came at the same time. He gave himself insolent airs and was evidently prepared to be very troublesome. He insisted that by the orders of the governor the mûnshi was to accompany us. After what had occurred we had no wish to have this person with us, and protested against it. He appeared to be equally unwilling to continue in our company I proposed, therefore, that we should go to the palace to have the matter settled. The soldier, however, declined to do so, and was so impertinent in his language that I was provoked into calling him a 'dog'—a very opprobrious term of abuse in Persia, although in common use—whereupon he jumped from his horse, and seizing a heavy stick, made a rush at me. I was prepared to meet him, and the quarrel might have ended disastrously to one of us had not the mûnshi thrown himself between us. I was determined to see the governor, but he was at the house of the Aga for whom I had brought letters from Baghdad, and who proved to be a fanatical and bigoted mulla. We endeavoured in vain to get into the house, the entrance to which was choked with the attendants and guards of the Serdar. The soldier, however, managed to reach him, and came back with a message that the mûnshi might either accompany us or not, as he thought fit.

As it was useless to attempt to force our way into the mulla's house, we determined to lose no further time, but to proceed at once upon our journey, threatening to complain to the Shah of the manner in which we had been treated. The soldier appeared to be alarmed at our menace, and disappeared. We, therefore, made our way out of the town without him. It was nearly sunset owing to these delays before we passed through the gates. On the top of a tower in the principal square was a band of musicians, who were producing the most discordant noises with drums of various sizes, and trumpets, some of which were of immense length, such as are seen in ancient Persian sculptures. This was a ceremony performed every day as the sun went down.

As we left the town we were joined by some of the Armenians who had formed part of our caravan from Baghdad, and who wished to continue their journey under our protection. We also met the governor's secretary, to whom we made our complaint against the soldier. He pretended to be surprised at his misconduct, and promised to report it to the Serdar. We were now a party of eight; but the roads between Kermanshah and Hamadan, where the Shah had established his camp, were reported to be safe in consequence of the troops which were constantly passing to and fro. We travelled through the night, reaching the celebrated rock-cut sculptures of Bisutun, or Behistun, before sunrise, and established ourselves for the day in a large khan. The soldier who was to have accompanied us to the royal camp had not made his appearance, and we were left again to ourselves.

We passed some hours in examining the celebrated bas-reliefs and cuneiform inscription carved on the scarped rock by King Darius.[6] But they were at so great a height from the ground, and so completely inaccessible, that it was impossible to make copies of them.

We travelled again during the night, the heat, although we were in a high mountain region, being still great. Our journey, although of only five farsaks, or about fifteen miles, was a very fatiguing one, as we lost ourselves in marshes, in

[6] This inscription in three columns and three languages has, as is well known, furnished the most important materials for the decipherment of the cuneiform character. Sir Henry Rawlinson, some years after my visit to Bisutun, was able, by having proper means of access to them, to make copies, and subsequently published his translation of the Persian column which contains the history of the Great King. The two other columns contain the same inscription in the Babylonian and Median languages.

which our horses were constantly sinking. The croaking of innumerable frogs almost deafened us. At last we reached the village of Sahannah, or Sanna, where we spent the day. It was July 8 —just one year since I had left England on my adventurous journey. I was still in high spirits, deeply interested in all that I was seeing, enjoying to the full my independent mode of life, ready to face any further perils and difficulties, and in excellent health, notwithstanding the attacks of intermittent fever from which I had occasionally suffered, but which did not appear to have produced any permanent effect upon my naturally robust constitution.

Early next morning we arrived at Kangowar, a large town situated on a hill-side. We found that the Shah had moved his camp to this place, and that every house and khan was occupied by his attendants and by public functionaries and their followers. We could find no accommodation, and were obliged to put up with a wretched stable swarming with vermin, which we shared with our horses. The troops had pitched their tents in a small plain near the town, and in the extensive gardens which surrounded it. Long strings of horses and mules laden with provisions for the army were constantly arriving, and horsemen were galloping to and fro in all directions. The scene was a very animated and amusing one. The Shah himself arrived soon after us, followed by a large body of troops and by a great retinue, and greeted by a wild and barbarous music of trumpets and drums. He rode to a house in one of the gardens which had been prepared for him. We could not ascertain the number of his troops, but we were assured, with the usual Persian exaggeration, that they amounted to 400,000 men, with 100 guns! There were probably about 13,000, with 16 guns. But there was a vast crowd of camp followers and of the idle and vicious rabble which usually accompanies an Eastern army.

I had been furnished at Baghdad with letters for Mirza Aga Baba, the Shah's 'hakim-bashi,' or principal physician, at that time an influential and much-respected personage at the Persian Court. In the afternoon, when something like order had been established in the camp, we found our way to the tent of this gentleman, who received us with the greatest kindness and courtesy. Twenty years before he had been in England, where he spent five years, and, I believe, married an English wife. He spoke our language with fluency. We informed him of our position, and of our object in passing through Persia, and asked his advice as to the course we should pursue. He recommended us to call without delay upon the Minister for

Foreign Affairs, to show him our passports and such papers establishing our character as simple travellers as we might possess, and to ask for a royal firman to enable us to continue our journey. We accordingly made our way to the minister's tent—a magnificent pavilion, lined with the finest Cashmere shawls and spread with the choicest carpets. Mirza Ali, who was at that time charged with the administration of the foreign affairs of his country, was a beardless youth of about two-and-twenty. In this important office he was associated with his father, Mirza Masoud, a statesman of experience and reputation, who was then absent on important public business in the province of Khorasan. He spoke French, and had among his secretaries a Frenchman. We had consequently no difficulty in making ourselves understood. He received us with politeness, looked at our passports, seemed satisfied with what we told him as to the object of our journey, and promised to speak on the subject to the Shah in the evening, and to send us his Majesty's answer next morning. He gave us to understand that no obstacles would be thrown in our way when travelling through his Majesty's dominions, and that we were at liberty to remain in the camp without being under any restraint.

We spent the remainder of the day in examining some ruins, apparently of the Greek period, consisting of the foundations of a vast building, constructed of enormous blocks of dressed stone. Eight columns still erect, and a pilaster, were half buried in the mud walls of a house in the town. They appeared to be the remains of a Greek temple, resting upon an earlier edifice, attributed, by one of those traditions so prevalent in this part of Asia, to Semiramis. Kangowar is supposed to represent the ancient city of Pancobar, where the Assyrian queen is said to have erected a temple to Anaïtis, or Artemis, and to have established an erotic cult in which, if her reputation be not belied, she was amongst the most ardent worshippers. We could not discover any inscriptions, but we heard from a native that there was a slab with strange letters upon it in a mosque. As we could not, being infidels, enter this sacred building, we sent Saleh, the Lur whose acquaintance we had made at Kermanshah, and who had followed us from that place, to report upon the alleged inscription, furnishing him with a pencil and a piece of paper, and directing him to make as exact a copy as he was able of any writing that he might discover. He returned shortly afterwards with a scrawl, which, however, was sufficient to show that an inscription did exist, and that the letters were apparently Greek.

The Shah left Kangowar at sunset on July 10, and the camp was immediately raised. It was an exciting and busy scene. The tents were thrown down, the baggage animals received their loads, the troops were assembled for the march. the irregular horsemen were careering over the plain, and Persians, with Lurs, Kurds, and other wild tribesmen, were engaged in mimic fight. The ministers and great officers of state, followed by their numerous retainers and servants, hastened after the Shah. In a short time this motley crowd had moved off, in the most complete disorder: not even the so-called 'ser-baz,'[7] or regular troops, maintained any discipline, but were scattered in detached groups along the road. They were a ragged and disorderly mob, with clothes in tatters and almost shoeless. Such at that time was the Persian regular army. It would not have held its own even against the ill-disciplined and disorganised troops of the Sultan.

The Shah was returning towards Tehran, having been induced by foreign pressure to give up the expedition against Baghdad, for which, it was generally reported, he had brought together his army and had approached the Turkish frontier. He had grievances against the Porte on account of the alleged ill-treatment of his subjects who yearly,

[7] Literally 'playing with life.'

in vast numbers, performed the pilgrimage, almost obligatory upon Musulmans of the Shi'a sect, to Kerbela and other holy places in Mesopotamia. He demanded, moreover, the restitution of the district of Zohab and other territory, which, he maintained, had been wrongfully occupied by the Turks. War had only, I believe, been prevented by the interposition of Russia, and by the fear that England might interfere on behalf of Turkey.

The plain, which an hour before had been covered with tents and resonant with the clamour of human voices, was now as silent and lonely as the desert. The greedy and undisciplined soldiers had not departed without leaving traces of their passage. Like a swarm of locusts, they had eaten up and destroyed almost everything that came within their reach. The vines had been rooted up and the fruit trees cut down for firewood; the standing corn had either been trampled under foot or carried away as food for the horses; the bazars and private houses had been pillaged of their contents. The wretched inhabitants of the town whose provisions had thus been consumed, and whose property had been wantonly devastated, would have reason to remember for many a day to come the visit of their sovereign and his army.

We followed the straggling and struggling crowd, which occupied a wide extent of ground on its march and rendered progress slow and difficult. In the middle of the day the Shah had encamped near the small village of Saadabad, and the tents of his troops and camp followers were pitched in its gardens. We had been travelling through a wide, well-watered plain, abounding in villages, which, from the trees and vegetation surrounding them, appeared from a distance to be prosperous. But they were all more or less deserted. Bad government and the repeated visits of tax-gatherers and soldiers had reduced them to ruin. We found a group of trees under which we could spread our carpets and enjoy the shade without being molested. I left my companion to keep watch over our horses and little property, and went in search of the tent of the Minister for Foreign Affairs, in the hope of hearing something about our promised firman, but was unable to find it. I came, however, upon that of Mirza Aga Baba, the hakim-bashi, who received me very kindly, and presented me to one Mirza Riza, an officer of engineers, who had studied in England and spoke English. He was then Director-General of the Shah's arsenals and foundries. I found him to be an intelligent and enlightened man. He told me that he had translated several English works into Persian for Prince Abbas Mirza, amongst them Sir Walter Scott's 'Life of Napoleon,' and a part of

Gibbon's 'Rise and Fall of the Roman Empire.' In the evening the hakim-bashi sent us a present of a lamb and a quantity of bread. The latter was very acceptable, as bread was very difficult to procure; but we had no means of cooking the lamb, and were obliged, at the risk of being considered rude and uncivil, to return it with our thanks. We were content with our pilau of boiled rice, but we suffered a good deal from the want of drinking water, which could only be procured from a considerable distance and for which we could not send.

The Shah remained during the following day at Saadabad, but the baggage and a great part of the camp followers were sent on to Hamadan, from which we were not far distant. We were now approaching the loftiest part of the great range of the Luristan Mountains, and the highest peaks were still covered with snow.

The troops were again on the march at sunset; but we remained until the middle of the night to avoid the crush in crossing the mountains of Elwend. We began a very steep ascent immediately after leaving the village. The night was dark; the track—it could not be called a road—execrable. It was still choked with stragglers and strings of baggage-horses, mules, and camels. We had no little difficulty in making our way. Accidents were constantly occurring, and the path was blocked by fallen animals and by drivers endeavouring to replace the loads. We were not sorry to find ourselves at sunrise in a plain on the other side of the pass, and to rest in a small grove of trees, not far from the place where the troops had encamped.

The Shah himself did not leave Saadabad until daylight. He passed close to us some three hours after we had alighted. We had a good opportunity of seeing him. He was preceded by the ladies of his harem, and by a number of women, enveloped in their thick veils and long garments, some riding on horseback, others carried in closed litters. Mohammed Shah, who rode a magnificent white Turcoman horse of great size, adorned with gold and silver ornaments, was accompanied by his son, a handsome boy of nine or ten years of age, and by his vizier, the Haji Mirza Agasi. He wore the usual Persian dress—the outer coat being of the most precious Cashmere shawl—with armlets of brilliants and an aigrette of diamonds in his black lambskin cap. He was followed by his ministers, his household, and a great retinue of officers and notables. Four elephants fantastically painted with all the colours of the rainbow, and covered with richly embroidered trappings, had been sent out from Hamadan to meet him. They formed part of the procession, which was closed by a body of irregular cavalry, comprising horsemen from the various tribes in his Majesty's dominions, who pressed onwards in the most indescribable confusion, amid clouds of dust.

After the Shah had passed we remounted our horses and followed him to the encampment. I succeeded in seeing the Minister for Foreign Affairs, who promised me a firman which would enable us to continue our journey as soon as his Majesty had reached Hamadan. Relying upon this promise we rode to the city, where we procured a room in a dirty, half-ruined khan. His Majesty arrived on the following morning.

Hamadan is a large and important place situated at the eastern extremity of an extensive and thickly peopled plain. The rugged mountains of Luristan rise on all sides in the distance, and to the south towers the lofty snow-capped peak of Elwend. It is abundantly supplied with water, which is led in open conduits through the streets and serves to irrigate a large tract of land, which is consequently clothed with perpetual verdure. Few cities in Persia have finer gardens and none is more renowned for its fruit. It is surrounded by trees, among which the poplar is conspicuous, and has from a distance a very flourishing and pleasing appearance. The bazars were extensive, and well supplied with produce and merchandise of various kinds; but they were in a ruinous condition, as were the numerous khans for the reception of travellers and of merchants with their goods. The houses, like those of other Persian, and, indeed, of most Eastern, cities, had no exterior architectural ornamentation. The outer walls, built of sun-dried or baked bricks, were without windows, and had only one doorway, from which a dark vaulted passage, just large enough to admit a horse, led into the interior.[8] Within, however, the dwellings of the principal inhabitants of Hamadan were remarkable for the beauty and richness of their decorations. In the centre of the courtyards, upon which the apartments of the men and of the women opened, were fountains of sparkling water and parterres of gaudy flowers. The rooms were painted with the most intricate and graceful designs, in brilliant colours, and profusely gilt. The coved ceilings were ornamented with numberless little mirrors arranged in patterns, which reflected the objects below, and produced, especially when the room was lighted after dark, a most enchanting and fairy-like effect. The Iwan, or hall in which the owner received his guests, was panelled with a

[8] As the women in Persia are always kept closely concealed, the houses have generally no windows looking into the street which would allow them to see, or be seen, by men.

greyish marble, elaborately carved. The pavement, of the same material, was spread with carpets and felt rugs of the finest texture—the renowned produce of the handlooms of the Lur and Kurdish tribes. Water was led in marble channels through these halls, and in the centre of each was a fountain constantly playing, which gave a delicious coolness to the air and invited to sleep by its gentle plashing. It would, indeed, be difficult to imagine anything more truly enchanting than these abodes of the nobles and of the wealthy merchants of Hamadan.

The mud-built houses, or rather hovels, inhabited by the poorer classes were mostly in ruins. The streets were narrow and unpaved, and deep in mud and filth. The city contained numerous Armenian families, who occupied a quarter of their own distinct from that of the Musulmans. Owing to the vicinity of the mountains, the great elevation above the sea of the plain on which Hamadan is built, and the abundance of water which flows in continuous streams within and about it, its climate is proverbially healthy. The air is cool and agreeable in summer and not too keen in winter.

We had made the acquaintance of Monsieur Nicholas, a French gentleman acting as a secretary to the Minister for Foreign Affairs. Seeing us so ill lodged in our dirty and ruined khan, he kindly proposed that we should share the quarters which had been assigned to him. We gladly accepted his invitation, and found ourselves in a fairly clean and habitable house, where we were at least free from the vermin with which the caravanserais in Persia abound. We were detained for nearly one month in Hamadan, owing to the difficulties we experienced in obtaining our firman and permission to continue our journey through Persia. But our time did not pass unpleasantly or unprofitably in the society of Monsieur Nicholas and some French officers then in the service of the Shah as military instructors. They were well-educated men, and possessed a few books which they kindly lent me. They were, moreover, able, from the experience they had gained by a residence in the country, to give me some useful information. I became also acquainted with a few Persian gentlemen, some of whom had been in England, like the hakim-bashi, spoke our language, and were consequently less intolerant and bigoted than the rest of their countrymen. They were courteous and obliging, and I passed a good deal of my time with them, increasing my knowledge of Persian, and preparing myself for my future travels by studying the manners of the people.

As soon as the Shah and his retinue had

established themselves in Hamadan, we called upon the Minister for Foreign Affairs and urgently requested that we might no longer be delayed, but that the promised firman might be given to us. We were promised it immediately. Soon after we were taken by Monsieur Nicholas to visit a Persian nobleman named Mahmoud Khan, who had married one of the Shah's sisters, and who resided in a village about two miles distant. He was a handsome and very intelligent young man. Although he had never been in Europe and spoke no European language, he was above the usual prejudices of his countrymen. He gave us an excellent breakfast, of which he partook with us, and allowed us to use his kaleóns. His country-house, a very handsome and richly-decorated building, but in a neglected and somewhat dilapidated condition, stood in the midst of a garden abounding in sweet-smelling flowers and watered by innumerable rills of clear running water, in both of which Persians delight. To spend their time idly amongst them, stretched on a carpet spread upon the grass, smoking kaleóns, listening to music and the nasal drawl of the reciters of the verses of their favourite poets, and watching the tortuous movements of dancing boys or, when possible, of dancing girls, and swallowing glasses of fiery 'arak,' appeared to them supreme happiness.

Our host indulged in these pleasures, and we passed the afternoon in their enjoyment, without, however, partaking too freely of the vile spirit which was constantly handed round by his attendants—slipshod youths in gay-coloured flowing dresses, with the nails of their fingers and toes and their locks dyed with henna, and with one hand upon the jewelled haft of the 'hanjar,' or curved dagger, stuck in the Cashmere shawl encircling their waists. It was the first time that I had seen a Persian orgy. Although the Khan was rather unsteady on his legs from the quantity of 'arak' he had drunk, he was able to accompany us back to the city on horseback.

Next day we again called on the Minister for Foreign Affairs, who informed us that he had been commanded to present us on the following morning to the Shah and to the Prime Minister. When the time came, however, his Majesty was unable from indisposition to receive us, and although he subsequently expressed a desire to see us, we pleaded the want of a proper dress in which we could appear before him in order to avoid an audience. But we waited upon the Prime Minister, the Haji Mirza Agasi, who was then the man of the greatest influence, power, and authority in Persia. The Shah had committed to him almost the entire government of his kingdom, occupying himself but little with public affairs, aware of his own incapacity for conducting them. 'The Haji'—the name by which he was familiarly known—was, by all accounts, a statesman of craft and cunning, but of limited abilities He was cruel and treacherous, proud and overbearing, although he affected the humility of a pious mulla who had performed the pilgrimage to Mecca and the holy shrines of the Imaums. The religious character which he had assumed made him intolerant and bigoted, and he was known to be a fanatical hater of Christians. He had been the Shah's tutor and instructor in the Koran, and had acquired a great influence over his pupil, who had raised him to the lofty position which he then held. He had the reputation of being an accomplished Persian and Arabic scholar, but he was entirely ignorant of all European languages. His misgovernment, and the corruption and general oppression which everywhere existed, had brought Persia to the verge of ruin. Distress, misery, and discontent prevailed to an extent previously unknown. He was universally execrated as the cause of the misfortunes and misery from which the people and the State were suffering.

We found him seated on his hams, in the Persian fashion, on a fine Kurdish carpet spread in a handsome hall. Before him was a large tray filled with ices and a variety of fruits. He was a man of small stature, with sharp and somewhat mean and forbidding features, and a loud shrill voice. His dress was simple—almost shabby—as became a mulla and a man devoted to a religious life.

He received us civilly, welcomed us to Persia, and questioned us as to the objects of our journey. We informed him of our intention of reaching India through Yezd and the Seistan. Why, he asked, had we remained so long at Baghdad? We endeavoured to satisfy his curiosity. The shortest and easiest road from that city to India, he said, was by Bushire and the Persian Gulf to Bombay, and not by land through Persia. We explained to him that we were fond of travelling, and that we were desirous of visiting the dominions of the Shah, and of examining the site of ancient cities in parts of Asia which had not hitherto been explored by European travellers. Mr. Mitford added that he suffered so much when at sea that he preferred the longest and most difficult land journey to a voyage in a ship. 'How, then,' he exclaimed, 'could you travel on the plains of Baghdad, which are known to be excessively damp?'

He then informed us that we could not be permitted to pass through Yezd and the Seistan, as the roads through that part of the Shah's territories were then in a dangerous state, and if any disaster happened to us he would be held answerable for it by the British Government. An accident, he said, to an English courier on his way to Herat had already led to differences between England and Persia, which had ended in the withdrawal of the British Ambassador from Tehran. If we wished to go to India there were two roads open to us—either that by Shiraz, Bushire, and the Persian Gulf, or that through the north of Persia, Herat, and Afghanistan. We might choose either of them, and we should receive all necessary assistance and protection for the prosecution of our journey. But he would not, he declared, allow us to pass through the Seistan.

I attempted to shake his resolution by pointing out that there was no similarity between our case and that of a courier who was employed by the British Government on public service. We were simple travellers, who were acting entirely upon our own responsibility. If any accident happened to us after the warning he had given us, neither we nor any one else would have a right to complain. I cited the case of Captain Grant and Lieutenant Fotheringham, who had been murdered by the Lurs, and for whose death the British Government had not considered it necessary to demand redress. Now that our representative had been withdrawn, and the relations between the two countries were suspended, we could no longer appeal to our Government for peotection. As we had ventured into a country with which England might be considered in a state of war, if any harm came to us our blood would be upon our own heads.

He appeared to yield to these arguments, and offered to allow us to proceed to the Seistan through Yezd if we would sign a declaration to the effect that we had taken this route contrary to his advice and in spite of his warning, and that he was consequently in no way responsible for our safety. This I willingly consented to do; but Mr. Mitford hesitated to agree to the proposal, as he justly thought that such a declaration might encourage an attempt upon our lives if there were any intention or desire on the part of the Persian authorities to get rid of us. It was evident that the Haji suspected that we were spies and agents of the British Government. However, he declared that the Shah was willing that we should visit any part of his territories where we could travel in safety, and that orders had been issued for the preparation of our firman; for his Majesty had said that we belonged to a friendly nation, and his quarrel was not with England, but with Lord Palmerston. who had treated Persia ill, and had recalled the Queen's Ambassador without sufficient cause. He would, therefore, make no difference in his treatment of Englishmen, and we should enjoy his protection and receive every assistance during the time we were in his dominions. These were the very words, the Haji declared, of his royal master, and he added many civil things to them. But he changed his mind as to allowing us to pass through Yezd, as the country between that place and the Seistan was even in a more disorganised condition than usual, on account of the occupation by England of Afghanistan. There was, moreover, a report that an English traveller had been murdered in attempting to reach the Lake of Furrah, which it was one of our objects to explore.[9]

I then proposed that we should be allowed to pass through Kerman and Beloochistan. But to this he would not consent, alleging that the danger by this route would be as great as by that through the Seistan. It was useless to argue further with him, and our interview ended. Mirza Aga Baba, whom we afterwards saw, and who was disposed to be very friendly to us and to give us all the help in his power, confirmed our suspicions that the Prime Minister had taken us for spies, and told us that he was convinced that we should not be permitted to pass either through Yezd or Kerman, as

---

[9] The report proved to be true. The traveller murdered was Dr. Forbes.

it was believed that in visiting that part of the country our real object was to explore a new route by which an English army could be sent from Afghanistan to invade Persia. We had, therefore, to choose between the northern route through Meshed and Herat, and that by Shiraz, Bushire, and the Persian Gulf. We were inclined to take the former, in the hope that once on the Herat frontier, beyond the jurisdiction of the Persian authorities, we might be able to carry out our original plan of visiting the Seistan and the Lake of Furrah, and of tracing the course of the river Helmund, which was believed to fall into it.

The hakim-bashi corroborated the Prime Minister's statement that the Shah had expressed himself in a very friendly manner with regard to us, and that he had given orders that a 'mehmandar,' an officer charged with the care of travellers, and especially those of rank, was to accompany us. Moreover, the Haji was willing, he said, that we should travel at the Shah's expense if we would only quit the Persian territories without delay. I had informed Mirza Aga Baba that I was anxious to visit some ruins supposed to be those of Shushan the Palace, which, there was reason to believe, were of great interest, and that to reach them it would be necessary for me to pass through the mountains inhabited by the Bakhtiyari, who had the reputation of being the most savage and lawless tribe in the Shah's dominions. He quite understood my curiosity, and was too intelligent to suspect that I had any other object in view but to gratify it. He said that the Haji could not admit that any part of Luristan was in rebellion to the Shah, and he undertook to obtain for me permission to cross the mountains to Shuster, a city in the vicinity of the site of the ancient Susa, and to recross them to Isfahan.

Hussein Khan, who had recently returned from London, where he had been sent by the Shah as his ambassador after the rupture with the British Government, was present during our interview with the Haji. He had not been received officially in England, but he spoke warmly of the hospitality and kindness that he had experienced during his residence there. As he asked us to call upon him we did so. He was profuse in civilities, and offered to obtain permission for us to visit Yezd and Kerman if we would undertake not to proceed beyond those places, but to return to Isfahan. This I declined to do, as I had no other object in going there but to make my way through the Seistan to Kandahar. I found afterwards that Hussein Khan bore a very bad character. He was accused of having appropriated to his own use the pay and allowances of several French officers whom he had induced, when at Paris, where he was also sent as ambassador, to enter the Shah's service for the purpose of instructing the Persian troops and organising a regular army. He charged the Government for their travelling expenses at a high rate, although he had compelled the inhabitants of the towns and villages through which he had passed with them to supply gratuitously both provisions and carriage. Finding himself greatly in debt on his return from his mission, through his extravagance during his journey in Europe and his residence in Paris and London, and the superintendent of his estates not being able to furnish him with the money he required, he accused him of having embezzled it. He placed the unfortunate man in confinement and inflicted the most cruel tortures upon him, even compelling the son of his victim, a boy of only six years old, to burn his father with hot irons, and giving his wife over to the 'farrashes,' or common servants. He died under the treatment to which he was subjected. The mullas of Tabreez, where these atrocious acts had been committed, were, as pious Musulmans, horrified by the outrage upon a Mohammedan woman, and addressed a petition to the Shah demanding that Hussein Khan should be punished with death for committing a crime considered worthy of it by the law of Islam. However, by dint of large presents to the Shah and the Haji, he escaped the punishment he so fully deserved. I heard afterwards that when his Majesty learnt that his late ambassador had swindled him out of a considerable sum of money in the matter of the French officers, he ordered him to be bastinadoed on the soles of his feet. The strokes were administered so effectually that he lost the nails of his toes, and was unable to walk for many weeks afterwards. He lost also, for the time, his royal master's favour.

It was evident from the information we had received that we should encounter very great difficulties in attempting to pass through the Seistan to Kandahar. The Persian Government were resolved to prevent us from doing so, and if we ventured to proceed without its authority and in spite of its opposition, trusting to our disguise, our lives would be in imminent danger in a country notorious for the lawlessness of its inhabitants. Mr. Mitford was unwilling to incur the risk, and being now anxious to reach his destination, determined upon taking the most direct route—that through the north of Persia by Meshed and Herat. But it was not without its dangers, owing to the disturbed state of Central Asia. He was, however, assured that he need be under no fear so long as he was within the Shah's dominions, and he hoped on arriving on the Afghan frontier to be able to communicate with the British authorities at Kandahar, and with their assistance to reach that place.

I was unwilling to renounce the attempt to reach the Lake of Furrah. I was not without hope that at Isfahan I might find an opportunity of joining a caravan, or a party of travellers, going to Yezd, and that I might even perform the journey without attracting the notice of the Persian authorities. I determined, therefore, to separate from Mr. Mitford and to proceed in the first instance to that city. We accordingly asked for separate firmans, which were promised to us. But we soon learnt the value of Persian promises. It was not until August 8, after having been detained for nearly one month at Hamadan, that we obtained the documents we required and the permission of the Shah to continue our journey. We spent the greater part of that time in going backwards and forwards from the Prime Minister to the Minister for Foreign Affairs. We were always received with politeness, our remonstrances were listened to, and we were assured that on the following morning, without fail, we should be in possession of all that was required to enable us to take our departure. The morning came, but not the firmans. We were the more anxious to leave Hamadan as in riding through the town and the camp we were exposed to constant annoyance and insult, and were occasionally in some danger. The population of the city was fanatical, the soldiers were insolent and without discipline, and there were in the irregular cavalry wild fellows from the mountain tribes, who would not have scrupled to take the life of a Christian and a European. Stones were frequently thrown at us as we rode among the tents. We were occasionally threatened with actual violence, and in the streets we were usually greeted with cries of 'Káfir' (infidel), 'dog,' and other opprobrious epithets. This state of things was only partly put an end to when, on one occasion, a sentry having hurled a large stone at me which struck my horse, I proceeded to the Haji and demanded redress, threatening to appeal to the Shah himself unless it was afforded me. I was able to identify the culprit, who was arrested and received a bastinado. We were not afterwards molested in the camp, but in the city we were constantly insulted in the most foul language.

Fortunately for us the Baron de Bode, who was then First Secretary to the Russian Embassy in Persia, arrived at Hamadan on a special mission to the Shah. The name of this gentleman, who, although in the service of the Russian Government, was, I believe, partly of English and partly of German descent, is well known from the large pecuniary claims of his family upon the British Government which were frequently brought before the House of Commons, but which were never established. We called upon him, and were received with great courtesy. He promised to speak to the Prime Minister in our behalf, and to remove from his mind the suspicions that he entertained with respect to the object of our journey. He kept his word, and it was probably owing to his assistance that we at last obtained our firmans. We saw a good deal of him during our detention, and found much pleasure in his society. He was a well-informed and accomplished man, interested in geographical and archæological subjects, and ready to assist those who were engaged in the same pursuits. He also kindly advanced me a small sum of money upon a draft on London. I had been robbed of my purse, which contained almost all my available funds, and was without the means of continuing my journey.

Hamadan is known to occupy the site of Ecbatana, the ancient capital of the Medes. I explored the city and its neighbourhood in search of ruins, but without much success. A few mounds on an eminence at a short distance from the walls may mark the site of an ancient castle or palace, but with the exception of the shafts of some marble columns, and the figure of a lion rudely sculptured in stone, I found nothing to reward my trouble. Yet Ecbatana appears to have been a city scarcely inferior in size and importance—if the accounts of early Greek writers are to be trusted— to Babylon. It was celebrated from the remotest times for its wealth, for its walls covered with plates of gold, and for the enormous strength of its fortifications. Its foundation was attributed to Semiramis, who adorned it with a magnificent palace and many temples. It was renowned for the seven concentric walls of different colours, by which, according to Herodotus, it was surrounded, to represent the seven heavenly bodies. That such walls ever existed is more than doubtful. I could find no traces of them. The city must undoubtedly have been one of considerable size as well as of great antiquity. Arbaces is said to have made it his capital after the fall of Nineveh; and the 'Great King,' according to Xenophon, was accustomed to pass the summer months in its cool and delightful climate. Here Alexander the Great, on his return from the far East, stopped and offered up sacrifices to the gods. Here also his favourite, Hephæstion, died. It remained a place of considerable importance during the time that the Parthian and Sassanian dynasties held sway in

Persia, as it stood on the great highway which from the earliest times led from Babylonia, over the lofty mountains of Zagros, into Media. Coins, gems, and other objects of antiquity are constantly discovered when the soil is turned up within and around the city. Hamadan has become a well-known mart for such things.

According to a Jewish tradition, Esther and Mordecai died in Ecbatana. A modern building, surmounted by a cupola, has been built over the place in which they are said to have been buried. The spot is held in the greatest veneration by the Jews, who flock to it as a place of pilgrimage at certain periods of the year. As is usually the case when a place is held sacred by either Christians or Jews, the Mohammedans have claimed it as a shrine of one of their own saints, and are also accustomed to make pilgrimages to it I found nothing in the building except a vault filled with rubbish. Innumerable bits of rag had been fastened to the walls of the tomb by pilgrims who had visited it for devotion.

Hamadan is also said by Arab writers to contain the tomb of the celebrated Aben Sina, more generally known by the corrupted form of his name as Avicenna, the renowned Arab philosopher and 'Prince of Physicians.' I searched for it in vain. Several persons informed me that it still existed in the city, but no one seemed disposed to take me to it. Being probably looked upon as a sacred spot, it was not considered right by the fanatical population that it should be polluted by the presence of an infidel.

Although in the city itself there are scarcely any ancient remains, there exist at about three miles distant from it, in the mountains near the village of Abbasabad, some very important inscriptions in the cuneiform character. I rode to the place through a valley, under the grateful shade of wide-spreading trees and by the side of a clear, rapid stream. On either hand were orchards rich in various kinds of fruit—peaches, apricots, plums, and melons. Vines laden with grapes covered the hill-sides. In the distance rose the majestic peak of Mount Elwend. It was altogether an enchanting scene, the more delightful from the contrast it afforded to the crowded and filthy city and camp.

The inscriptions occupy two tablets about six and a half feet in height and eight and a half in breadth, cut in a rock or cliff closing the end of a narrow gorge, through which flows a stream having its source immediately beneath them. Other tablets have been prepared for similar records, but which, for some reason, have not been used. The two that have been completed contain each three inscriptions placed in parallel columns. They are trilingual, and in three forms of the cuneiform character, the wedge being the element in all three, but differently arranged to form letters and signs It is now well known that they represent the three different languages spoken in the vast dominions of the ancient Persian kings, of the so-called 'Kayanian dynasty'—the Persian, the Babylonian, and the Median. The inscriptions contain the names and titles of Darius Hystaspes and of his son Xerxes, with invocations to Ormuzd, the supreme deity. They are of special interest, as having first afforded the key to the decipherment of the cuneiform writing.

It took me three hours to make as careful a copy of the inscriptions as my then limited acquaintance with the character, the difficulty of access to one of the tablets, and the condition of their surface, which in many places had been worn by the effects of the weather, allowed me to obtain.

At length, after long and tedious delay and after constant applications and protests, we received our firmans, duly sealed by the Shah, on August 8. His Majesty treated us generously, and ordered that we were to travel at the public expense. We were to be furnished, without payment, with a certain number of horses. It was specified in my firman that I was to receive at every place where I stopped for the night provisions for eight persons, including chickens, meat, eggs, rice, bread, sugar, and many other things, and barley and straw for my horses. The 'mehmandar'[1] who was to accompany me and to see to all my wants, was to give receipts to the heads of the villages for the provisions supplied to me, the price of which was to be allowed to them in their taxes and other payments to the Shah's treasury. As I well knew that this was a mere idle form, that the villagers themselves would have to bear the expense, and that these rations and allowances to travellers of rank are made the excuse for great oppression and extortion, I determined not to avail myself of his Majesty's liberality, but to travel as economically as I possibly could and to pay for all I required. I should have gladly dispensed with the attendance of the mehmandar, but as he had received the Shah's express commands to accompany me, and was probably set to watch and report my movements—the Haji not having divested himself of the suspicion that I had other motives for travelling than those of pursuing geographical and antiquarian researches—I was compelled, very unwill-

---

[1] The 'mehmandar' is an officer appointed by the government to accompany travellers of distinction and to provide for their wants.

ingly, to retain him. The governors of districts and towns on my way were ordered to furnish me with escorts whenever danger was to be apprehended, and were made responsible for my safety.

In addition to the firman I received a letter from the Haji to Mehemet Taki Khan, the great Bakhtiyari chief, recommending me to his special protection. I was also furnished with a letter to the governor of Isfahan, who was directed to afford me facilities for the prosecution of my journey.

On August 8 I rode with Mr. Mitford as far as the village of Shaverin, where we dined with the French officers. I afterwards took leave of my companion, who started on his long journey through the north of Persia to Kandahar. We had been together for above a year, and I much regretted that we had to part. He had proved an excellent fellow-traveller, never complaining, ready to meet any difficulties or any hardships, and making the best of everything.

I then returned to Hamadan. The Shah had left in the morning and his camp had been raised. There was silence and desolation where a few hours before there had been tumult and bustling crowds Before leaving the city the soldiers had pillaged the bazars. All the shops were closed, and the inhabitants, dreading violence and ill-treatment, had concealed themselves in their houses. The gardens around the town had been stripped of their produce and the trees cut down. The place looked as if it had been taken and sacked in war. Such was the usual result of a visit from the Shah, his Ministers, and his army.

CHAPTER VII.

Leave Hamadan — My mehmandar — Douletabad — A Persian palace—Kala Khalifa—Burujird—Khosrauabad—Difficulties of the journey—A village chief—The Bakhtiyari—One of their chiefs — Renounce attempt to reach Shuster — Freydan—A Georgian colony—Tehrun—Reach Isfahan—M. Boré—Mr. Burgess—The Matamet—The bastinado—Imaum Verdi Beg—Shefi'a Khan—Ali Naghi Khan—Invitation to Kala Tul—Delays in departure—Residence at Isfahan—Messrs. Flandin and Coste —The Palaces—Persian orgies—The Mujtehed.

I WAS now alone. The most arduous and dangerous part of my journey to India, if I persisted in my attempt to reach Kandahar through the Seistan, was before me. In order to be entirely independent in my movements, and to be able to choose the route which suited me best, I had bought from a soldier a strong sturdy horse. As it had probably been stolen I paid but a few tomans[1] for it. All I possessed in the way of luggage was contained in a pair of small saddlebags. I was not, consequently, in need of a second horse for my baggage. My quilt and

[1] The 'toman' was then worth about 10s.

carpet were placed over my saddle. It would have been impossible to travel with fewer encumbrances.

It was not until the afternoon of August 9 that a Ghulâm,[2] named Imaum Verdi Beg, who had been appointed my mehmandar, had completed his preparations for the journey and was ready to start. We could, therefore, make but a short stage. He joined me, mounted on a good horse, and in travelling costume—his robes thrust into a huge pair of breeches of brown cloth—armed with a long gun, a huge pistol, and the usual curved dagger, and various contrivances for holding powder and balls hanging from his belt. We left the city together, and rode through a well-cultivated and fertile plain, thick with habitations surrounded by trees and gardens, and watered by numberless streams. In about two hours we reached the large village of Yalpand. The Ghulâm put his horse to a gallop when we came in sight of it, to precede and prepare a lodging for me. He secured a clean and airy room for me at the top of the best house in the place, and when the sun went down, an excellent supper with a variety of dishes was served to me.

When, in the morning, I wished to pay for my night's entertainment, I was informed that I

---

[2] The title given to an officer in the household of the Shah or of any great personage.

was the Shah's guest, and that, consequently, I was travelling at his Majesty's expense. I remonstrated in vain. The Ghulâm declared that the royal firman must be obeyed, and that no one would dare to receive money for anything supplied to me.

Our departure was delayed by a quarrel between my mehmandar and the head of the village. I then discovered that he had sent back his horse to Hamadan the previous evening, as he wished to spare it the long journey to Isfahan. He was now demanding from the villagers the horses with which, according to the Shah's firman, they were bound to provide me. After a great deal of wrangling and threatening, he succeeded in obtaining a wretched horse, and a donkey upon which a load was placed. What the load consisted of I could not at first imagine, as he had not been encumbered with luggage on his departure from Hamadan. I soon discovered that he had already commenced a system of extortion, for which the inhabitants of the villages at which I might stop for the night were to be the victims during the whole of my journey to Isfahan. My firman specified the supplies that I was to receive at each place. The Ghulâm had exacted them at Yalpand, and as they were far beyond what he or I could consume, he insisted upon carrying off the surplus. This accounted for the donkey's load I was very angry, declared that I would not be a party to so flagrant an abuse of the Shah's orders, and that, much as I valued his Majesty's generosity and hospitality, I would not profit by them to the detriment of his subjects. But the mehmandar persisted. He argued that as he had given a receipt for what he had taken to the 'Ket-Khudâ,' or head of the village, the inhabitants would be repaid from the royal treasury, and that if he had not exacted all the supplies granted to me they would nevertheless be charged to the Shah Why, therefore, should we not profit by his Majesty's bounty instead of the 'gourum-sags'—the scoundrels—who wished to cheat him? Although the argument had some weight, I could not reconcile myself to the idea of travelling at the public expense, especially as I was well aware that the villagers had but little chance of being repaid out of the Shah's empty treasury. I again protested that I was resolved to pay for all that had been supplied to me and my horse. But both the master of the house in which we had lodged, and the Ket Khudâ, were too much afraid of the consequences of offending a public officer to accept the money that I tendered to them, and I rode away in very ill-humour with my mehmandar, who was urging on the donkey, which, unable to keep up with the horses, greatly delayed our progress.

I had resolved to avoid the usual road between Hamadan and Isfahan, and to keep as close as I could to the great range of the Luristan Mountains. I should thus pass through a part of Persia which, I had reason to believe, had not been at that time explored by previous travellers, as it was a blank upon my map. After a pleasant ride through a hilly country abounding in villages, and offering constant views of high and picturesque peaks rising in the distance, we arrived late in the afternoon at Tashbandou (?), having breakfasted on our way, again at the public expense, at Samanabad. As the Khan, or chief, to whom the village belonged was absent, and the inhabitants did not appear inclined to obey the firman without his orders, I took up my quarters in the doorway of a small fort which he was constructing. In the meanwhile my Ghulâm was bullying and threatening the villagers, who, he declared, were 'yâghi,' or rebellious to the Shah, and, when reported as such to his Majesty, would receive condign punishment. He succeeded at last in finding a house, to which I removed.

During the night the man who was in charge of the horse and donkey carried off from Yalpand decamped with them, and we had fresh difficulties in procuring others. The Ghulâm, however, possessed, in addition to my firman, an order from the prince governor of Hamadan which entitled him to claim two horses at every village, or the amount of their hire in money. Although the inhabitants at first resisted the demand, they found the horses for him, after a few blows from the heavy whip of the officer administered right and left, and we continued our journey with a further addition to the supplies he had exacted at the other villages through which we had passed.

At the next village, Daëleh, where we stopped to breakfast, the inhabitants proved more loyal to the Shah, kissing the firman and pressing it to their foreheads, and supplied the required horses without delay. We continued through a hilly country, passing numerous villages and crossing many streams, and reached Douletabad early in the afternoon. I was surprised to find it a considerable town, although not indicated on the maps I possessed, surrounded by an embankment of earth and a ditch, and by double mud-built walls, the inner of which was very lofty and furnished with bastions. Passing through a gateway and through a heap of ruins, I found myself in a large quadrangle formed by low buildings having numerous arched recesses, serving for rooms, and handsomely decorated with stucco ornaments in relief. At one end of this square was an extensive palace, formerly the residence of the governor, but fast falling to decay. It must at one time have been a building of much magnificence. The walls of a spacious hall which I entered were painted in the brightest colours with human figures, animals of various kinds, birds and flowers, and arabesque ornaments.

Beyond this fine hall was a courtyard of large dimensions, in the centre of which was a tank of clear water, supplied by a spring. Around it were wide-spreading trees, rose-bushes, and flower-beds. At one extremity was a kind of screen concealing the entrance to an inner court, panelled with porcelain tiles of exquisite beauty, on which were enamelled in gorgeous colours the exploits of Rustem, the hero of the great Persian epic of the 'Shah-Nameh,' with numerous figures of warriors in mail and in fantastic costumes, and of horses with gaudy trappings.

I passed into this inner court, which was surrounded by numerous rooms partly in ruins, but still retaining remains of the ornaments in coloured stucco, glass, and carved woodwork with which they had been decorated. Beyond this court was a second, with fountains, rose-bushes, and parterres of flowers, and with similar rooms opening into it. It had been the enderun, or women's apartments.

I lingered with delight in admiration of these examples of Persian architecture and art in this deserted but still beautiful building until the Ghulâm, who had been to the governor of the town to obtain a lodging for me, returned. He had succeeded, and we left the palace together. We passed a fine mosque, the cupola and walls of which were covered with coloured tiles, and a kind of kiosk, in the form of a tower, elaborately painted, but falling to ruin. After making our way through a crowded and well-supplied bazar, we entered, through an archway, spacious pleasure-grounds intersected by avenues of lofty poplars, and watered by rills of running water, forming ponds and reservoirs. Roses and other flowers filled the air with a delightful perfume. Between the avenues were fruit trees and vines laden with grapes

In this garden were several detached kiosks, or summer-houses. One of them, standing on the margin of a little lake, had been assigned to me as a lodging. The room in which I spread my carpet was beautifully decorated with arabesques surrounding tablets on which were painted scenes from the chase—horsemen with spear and sword pursuing stags and hares, or more noble game, such as lions, tigers, and leopards; others with hawks on their wrists following partridges and other birds. In the centre of the room were two live falcons seated upon their perches, and in one of the corners were collected guns, swords, and spears. The palace and the kiosk, I learned, belonged to Prince Sheikh Ali Mirza, one of the sons of Feth-Ali Shah.

A large window, which could be closed with a wooden trellis of elegant design, opened upon a second garden with parterres of flowers and running water, even more spacious than that through which I had passed. Beyond was a long avenue of stately trees, which ended with a view of the cragged and snow-covered peaks of one of the mountains of the great Luristan range, called Kuh Arsenou.

I had scarcely seated myself on my carpet in this delicious retreat when two attendants placed before me an immense tray in which grapes, apricots, and other fruit were piled in pyramids. After I had eaten I wandered about the garden and entered one of the palaces, which was without inhabitants. It was a spacious building with a magnificent hall which, judging from the freshness of its coloured ornaments, appeared to have been recently restored. In the walls and ceiling small pieces of glass or mirrors were tastefully arranged in patterns—a favourite mode of decoration in Persia and Baghdad.

Other apartments which I entered were similarly decorated. Painted life size on the walls were figures of dancing girls in various postures, and of richly-clad ladies with almond-shaped eyes and black locks, as they are usually represented in Persian pictures, and hunting scenes, with horsemen bearing falcons on their wrists.

The Palace was reflected in a reservoir of crystal water, about a hundred paces in length. As I wandered through this beautiful building, which was without a human inmate and as silent as the grave, I might have fancied myself in one of those enchanted palaces whose inhabitants had been turned to marble, as described in the Arabian Nights, and which had so captivated my imagination in childhood.

It was not without much regret that I left this paradise, but time pressed and I could not stay. At five in the morning the governor sent a soldier to accompany us to a neighbouring village, where the Ghulâm expected that he would meet with difficulties in obtaining horses, for we were now approaching a country inhabited by a wild and lawless population little disposed to respect the Shah's firman. After leaving Douletabad we entered a highly cultivated and thickly populated plain. On all sides were villages, generally surrounded by mud-built bastioned walls, and containing a small fort in which the khan or village chief resided, as they were exposed to frequent attacks by marauding parties from the wild tribes inhabiting the mountains of Luristan. I rode through vineyards and fields white with the cotton-bearing plant. To the right rose the Elwend Mountains, which separate this rich plain— a blank on my map—from Luristan and the great range of Zagros Towering above them, and almost over-hanging Douletabad, rose the fine conical peak of Arsenou. After passing through a fortified village called Gouran, overlooked by a castle built upon a high and precipitous mound, we reached in about three hours Kala Khalîfa, where the Ghulâm stopped to procure fresh horses. The inhabitants at first absolutely refused to supply them, and it was only after a delay of nearly four hours that, with the aid of the soldier, he succeeded in obtaining a young horse and two donkeys to carry his increasing stores, exacted from the villagers as we went along. A Lur named Ali, who had accompanied us on foot from Hamadan, bought this horse for him for three tomans (thirty shillings). Imaum Verdi borrowed one toman from me, promising to repay it at the end of our journey. He sold the horse shortly afterwards for five shillings more than he gave for it, but did not offer to pay back the money I had lent him. In Kala Khalîfa there is a tomb said to be that of the son of the Imaum Ali, which is held in great veneration, and is a place of pilgrimage.

We now left the plain and entered the hills. They equally abounded in villages—each with its castle, its walls and bastions, having at a distance a rather imposing appearance, and showing the unsettled state of the country. The lands were irrigated by innumerable streams conveyed in artificial watercourses and in subterranean conduits called kanâts. A little before sunset we came in sight of Burujird, a large town situated in an extensive and well-cultivated plain, with the lofty range of Zagros, its higher peaks covered with snow, bounding it to the west. We did not, however, reach the gates until long after dark. I would not disturb the governor at so late an hour to obtain a lodging, but took up my quarters in a large and well-built caravanserai.

The Ghulâm, who had been sent to protect me, had already given me much trouble, and I had formed a very bad opinion of him. He now threatened in an insolent manner to leave me and to return to Hamadan, unless I gave him a sum of money far beyond what I could afford to pay I was not disposed to yield to his menaces, and told him that he might continue with me or leave me as he thought proper, but that in either case I should at once send a messenger to the Shah with a letter complaining of his behaviour. I reminded him, at the same time, of the fate of one Mirza Jaffer, the mehmandar of a French traveller, who, having been guilty of the same misconduct, had, upon complaint made to his Majesty, been condemned to lose his head. As he saw that I was resolved to resist the imposition, and was preparing to find my way to the governor to represent what had occurred and to engage a messenger to be sent to the Shah, he became alarmed, and implored me to pass over what had occurred and not to put my intention into execution. He went himself, at the same time, to the governor, to make, he said, all necessary arrangements for my journey to Korumabad. He returned shortly afterwards with many obliging messages from this official, and with assurances that I should be furnished even with fifty soldiers if they were needed to insure my safety in Luristan.

I remained at Burujird the next day in order to call upon the governor and to make arrangements to continue my journey. He was a Sirdar, or General, named Mirza Zamein. He received me at once, and expressed himself ready to help me as far as it might be in his power, but endeavoured to dissuade me from going to Korumabad. Not only, he declared, were the roads very unsafe on account of the unsettled state of the Lur tribes, who were in open revolt against the Shah, and were making constant depredations in the district through which I should have to pass, but the heat was so great at this time of the year at Shuster that no human being could possibly endure it. His statements were corroborated by others, and as I had reason to believe that there was some truth in them, I decided upon changing my route and endeavouring to reach Shuster by going to Freydan or Feridun, and thence to cross the Bakhtiyari Mountains.

To make up for his misconduct on the previous evening, the Ghulâm exerted himself to the utmost to please me, and I found him on returning to the caravanserai followed by several men bearing loads of provisions—bread, meat, fowls, rice, eggs, butter, tea, and firewood — enough to feed a regiment. He declared that they were a present from the governor, and that it would be considered a want of politeness and an offence on my part not to accept them. A very small portion of them sufficed for my wants; the rest went into the capacious sacks in which he had stowed the various supplies that he had been collecting on our way, and which he sold when he reached a town. In the afternoon I walked through the bazars, which I found extensive and well supplied with the produce of the country and foreign fabrics. The town, the position of which was wrongly marked on my map, contains about twenty thousand inhabitants, and is the largest in the province. It possesses several handsome mosques, whose domes and minarets give it a striking and picturesque appearance from a distance, and stands in the midst of extensive gardens and orchards, irrigated by streams coming from the hills. They are celebrated for their fruit, especially for melons and a small black grape of delicious flavour. In the bazar, melons, peaches, apricots, and plums were piled up in great heaps and were sold for a mere trifle. But this abundance of fruit is one of the causes of fevers and dysentery, from which the population suffer severely during the autumn. The town contains a few Jewish families, but no Christians.

I left Burujird early on the morning of August 14, and continued during the greater part of the day through the highly cultivated and thickly peopled plain which we had entered after crossing the hills of Douletabad. I had rarely seen a country so densely populated and with so prosperous and flourishing an appearance. We were evidently entering upon a district whose inhabitants had not been exposed to the oppressive rule of the Persian Government, with its attendant suffering and misery. It was harvest time in these high regions, and the peasants were everywhere engaged in cutting and carrying the corn. In all directions were long lines of beasts of burden, bearing sheaves of wheat and barley to the villages, where they were deposited on the threshing-floors, to be threshed by a rude roller, made of wood with iron spikes, drawn by oxen or horses. This mode of threshing prevails throughout the greater part of Western Asia.

About nightfall we stopped at the village of Khosrauabad. We were getting farther and farther from the country in which the authority of the Shah and his officers was fully recognised, and were entering upon that inhabited by the semi-independent tribes of Luristan. The Lur khan, the chief of Khosrauabad, declined to obey his Majesty's firman, and declared that he owed no allegiance to him. High words ensued. Imaum Verdi Beg drew his sword, and a very pretty quarrel, which might have led to bloodshed and serious consequences to myself, seemed to be impending. However, the khan at length yielded to alternate threatening and coaxing, and procured us a night's lodging. But the Ghulâm and our companion Ali were alarmed by these signs of rebellion, and declared that matters would get worse as we penetrated farther into the mountains of Luristan, where the authority of the Government was no longer recognised, and where it would consequently be impossible to obtain either provisions or horses. They urged me to give up the attempt to pass through the Lur Mountains to Shuster, and to take the direct road to Isfahan. To corroborate what they had told me about the dangers and difficulties of the route I proposed to take, they brought to me several 'charwardârs,' or muleteers, who were preparing to leave with a caravan for the latter city. They described to me, with circumstantial details, a number of murders and robberies which they affirmed had been recently committed upon travellers by the ferocious Lurs. As I could not depend upon the Ghulâm, and as it was evident that we had now entered a part of Persia in which the Shah's firman was no longer respected, I thought it advisable to make my way to the district of Freydan, instead of striking at once into the mountains. I hoped that thence I might still find means of carrying out my original intention of crossing the Luristan range to the plains of Khuzistan. If insurmountable difficulties were in the way I could always join the high road between Hamadan and Isfahan.

We had some trouble on the following morning in obtaining horses, but managed to resume our journey about seven o'clock. At the southern extremity of the plain of Burujird we entered a low range of barren hills. Although we still passed many villages they were not surrounded by gardens and trees as in the low country. But each had its small bastioned mud fort, generally perched upon a mound or a projecting rock, and having from a distance a very picturesque appearance. They are the residences of the khans to whom the villages belong, and serve as places of refuge for the inhabitants when they are engaged in the quarrels which constantly ensue between their chiefs, or when exposed to raids from the tribes of the neighbouring mountains. This part of Persia had always been in a very disturbed state, and its population appeared to live in perpetual warfare. Every petty chief considered himself independent of the Shah and at liberty to attack and plunder his neighbour, to carry off his corn, and to drive away his cattle. Life and property were nowhere safe, and the villages were for the most part in ruins. We saw in the distance during the day several encampments of black tents belonging to the Bakhtiyari, a nomad mountain tribe renowned for its courage and daring, and dreaded by the settled inhabitants of the plains. Their 'chapaws,' or forays for plundering villages and caravans, were carried on by bodies of horsemen to a great distance.

Even the neighbourhood of Isfahan was not safe from them. They were everywhere the terror of travellers and of the population.

We continued to skirt the lofty range of the Luristan Mountains, whose summits were covered with snow, which I was assured remained throughout the year. The names given to me for the principal peaks were Balighan and Shuterun. A river which we had hitherto been following now turned towards the south-west, and disappeared in a deep gorge, to issue again, I was told, in the plains of Khuzistan, or Susiana, near Dizful. I much regretted that I was unable to continue along it, and thus to reach by the shortest route the principal object of my journey, the ruins of Susan.

We stopped at the village fort of Miurudon (?), at the foot of Mount Shuterun. The khan was absent, but arrived soon after, accompanied by a crowd of ferocious-looking horsemen carrying matchlocks and armed to the teeth. He was a tall man, with a flowing black beard and a somewhat sinister countenance. He was probably returning from a raid, but he was civil to me, gave me a substantial breakfast, and asked me many questions about England, the Shah's army, which he heard I had seen, and my object in visiting his country. Although he professed to treat his Majesty's firman and the Ghulâm with the utmost contempt, he provided us with a horse and a couple of donkeys, and we continued our journey, reaching at nightfall the village of Derbend, the largest we had seen during the day, and, like those in the plain, surrounded by trees and gardens. I passed a sleepless night, assailed by myriads of mosquitoes.

We entered on the following morning upon a small plain in which were two villages, named Zarnou and Kirk, belonging to Armenians. It was divided by a range of low hills from a second plain inhabited by Bakhtiyari. The man who was in charge of the horses furnished to us at our last sleeping-place declared that he could not venture amongst these savage people, of whose deeds of murder and robbery he kept relating terrible instances, entreating us to see to our arms and to be prepared for the worst. He wished to take back the horses, and to leave the Ghulâm with his ill-acquired property to shift for himself. But Imaum Verdi Beg refused to part with them, as there were no others to be obtained. The poor fellow, seeing that we were determined to venture among the Bakhtiyari, and fearing to lose his life as well as his horses, took to his heels and left us in possession of them. They had been taken by force, and I was sorry for him, but there was nothing to be done. My mehmandar would not listen to my remonstrances, maintaining that it was only right that the Shah's firman should be obeyed.

In the extensive plain before us were numerous mud-built castles belonging to petty Bakhtiyari chiefs. We stopped at one of them named Makiabad (?). Najef Khan, its owner, welcomed me very cordially and invited me to share his breakfast, which was spread under a shady tree and consisted of 'âbi-dugh' (sour milk), a universal beverage in all parts of Persia, thick curds and cheese, with large cakes of unleavened bread. crisp and thin as a wafer, baked upon a concave iron plate over hot embers. He was a very handsome young man, with bright eyes and an open intelligent countenance. As we had the horses which had been left on our hands, there was no necessity for showing my firman, or of making any demand upon his village. We consequently parted good friends. I always used my firman unwillingly, and should not, indeed, have used it at all, except in cases of absolute necessity, as there would rarely have been any difficulty in obtaining the little I required from the villagers. But this document was unfortunately in the hands of the Ghulâm, who declared that as it had been confided to him by the Shah himself, he could not give it up to me until he had conducted me safely as far as he was ordered to accompany me. It was the source of constant trouble and vexation to me, as it led to quarrels and frequently to acts of violence on the part of Imaum Verdi on the poor villagers wherever we stopped on our road. When it did not suit him to carry off the provisions with which by the Shah's command I was to be furnished, he compelled the Ket-Khudâs to pay him their value in money. This led to continual protests on my part, and I had determined to lodge a complaint against him as soon as I reached Isfahan. When I warned him of my intention he sulked, although he was afraid to annoy me or to interfere with my movements.

Taking leave of Najef Khan, I resumed my journey through a hilly and barren country, thinly inhabited by Bakhtiyari. As the sun was setting I came in sight of what appeared to be a grand old castle on a mound rising above a village. It reminded me of one of those baronial strongholds of the Middle Ages of which ruins may yet be seen in many parts of Europe. These Bakhtiyari chiefs, indeed, lead the life of mediæval barons— at constant war with each other, plundering their neighbour's goods, his cattle and his flocks, and levying blackmail upon travellers and merchants However, as we approached, the illusion was soon dispelled. The village proved to be in ruins and uninhabited. The mud fort itself was scarcely in better condition. After riding with some difficulty up the steep ascent to it, I entered the gateway and found myself in a courtyard, in which were a number of armed men of very savage and sinister appearance lounging about. The khan soon made his appearance, and as fortunately there was no need to show my firman, and I presented myself as a simple wayfarer, he offered me at once a night's lodging and entertainment, and his followers were ready enough to help us and to see to our horses, for even the lawless Bakhtiyari, like all nomad tribes, consider themselves bound to receive a stranger and to treat the traveller with hospitality. The chief even offered to take charge of my saddle-bags, for the better security of their contents against thieves—an offer, however, which I thought it prudent to decline. The Ghulâm and Ali, our travelling companion, expressed great alarm at the aspect of the place and of its inhabitants. Before settling myself to sleep on my carpet I looked carefully to my arms, and prepared myself for any attempt that might be made upon my life or property in the night. However, our host had been apparently calumniated by the timid Persians, and I slept undisturbed.

I had been suffering for some days from a

severe attack of intermittent fever, and, in addition, from dysentery. As I felt very weak and scarcely fit to cross the Luristan Mountains by difficult tracks almost impassable to horses, where the population was as scant as it was hostile to strangers, and where I might find myself even unable to procure food, I decided upon proceeding at once to Isfahan, where I hoped to obtain some rest and medical advice before continuing my journey. I had been going through many hardships. The heat was still almost unbearable in the burning rays of an August sun, and I was obliged to travel during the day. My only bed had been for long but a small carpet, and I could never take off my clothes, which were in a very ragged condition. My food had consisted of little else than sour curds, cheese, and fruit. It was not surprising, consequently, that my health should have suffered.

We had now entered the district of Freydan, or Feridun, a considerable part of which belonged to the great Bakhtiyari chief, Mehemet Taki Khan. We stopped at the principal village in it, which bore the same name, and which contained about one hundred and fifty houses. It was inhabited by a Georgian colony, which had been established there by Shah Abbas. These Christians had retained their native language and their religion.

They were industrious, and their villages, which were numerous and surrounded by gardens and orchards, had a prosperous appearance. They were to be recognised at once by their features, which differed from those of the surrounding populations. Their women went unveiled, and many among those whom I saw were strikingly handsome. An abundance of water from the mountains, carried by innumerable watercourses and subterranean channels to all parts of the plain, irrigated a vast number of melon beds, producing fruit of excellent quality, which was sent for sale to Isfahan and elsewhere. A kind of clover, bearing a small fragrant flower, was also largely cultivated. We did not reach Adun, a Christian village where I had decided upon passing the night, until after dark. We were not hospitably received, and had much difficulty in getting a room. When at last we had succeeded in finding one, it was immediately crowded by idlers who came to gaze at the stranger, the news of whose arrival had spread through the place. Even the courtyard was filled with people who were waiting their turn to enter my room to stare at me. The women had congregated in numbers on the flat roof of the house, whence they could look down upon me through a hole in the ceiling which served for a chimney. I was placed to so much inconvenience by the men who crowded round me, that I was forced to threaten to drive them out with a thick stick. When at last they departed, I stationed Ali at the door, who kept guard with a drawn sword and would not allow any one to enter. The women could not be induced to withdraw, but remained on the roof watching my proceedings until I settled myself for the night. I had learnt by experience that in the East the Christians of all denominations were much less honest, hospitable, and considerate to a traveller than the Musulmans, and much less respectful and dignified in their manners. This may arise, as some maintain, from the inferior position which they hold, and from the ill-treatment they have experienced for so many generations from their Mohammedan rulers.

As I was now about to enter upon the track between Hamadan and Isfahan usually followed by caravans and travellers, I had no longer any need of the services or protection of a mehmandar. I had every reason to be dissatisfied with Imaum Verdi Bey. He had got me into constant trouble and quarrels in the villages by his extortions and the manner in which he was accustomed to treat the inhabitants. As the number of horses he required to carry the stock of provisions which he had been collecting during our journey could not be procured, he was obliged to be satisfied with donkeys. As these animals were unable to keep up with the horses, and were continually straggling into fields of ripe corn or barley to feed, my progress was much delayed and I lost a great deal of valuable time. Accordingly I insisted that he should deliver the firman to me, and I left him to do as he thought fit. He sold for ten shillings one of the donkeys which he had stolen, and then followed me. Ali came with him on one of the horses. We still skirted the lofty mountain range, from which rose a grand peak called Dulan-kuh, which had been visible during the previous two days. The plain through which we rode appeared to be deserted. We saw no villages, and the one or two caravanserais we passed were in ruins. The Ghulâm had been told that the inhabitants had fled on account of the incursions of the Bakhtiyari, and had been warned that we might probably fall in with one of their marauding parties. He was consequently very anxious that I should take an escort for my protection, which, however, I refused to do.

Towards evening we reached a small Bakhtiyari village, where we were unable to obtain either provisions for ourselves or barley for our horses. I saw a castle on a mound in the distance and galloped to it. But the place seemed deserted, and when I entered the gateway I found myself amidst a heap of ruins tenanted by a solitary herdsman with a pair of oxen. He could not help us, but said that there was a village off the road near the foot of the mountain where we might obtain what we required. It was already dark, but there was nothing to be done but to take the direction he pointed out to us. We were overtaken by a violent thunderstorm, and I soon got wet to the skin. Except when the vivid flashes of lightning, accompanied by deafening peals of thunder, showed us surrounding objects, we were in total darkness. When the storm had ceased and we had wandered about for some time, distant lights and the barking of dogs directed us to the village of which we were in search. After scrambling through ditches and wading through watercourses, we found ourselves at the gate of a ruined khan where some men were gathered round a bright fire. They were strolling shoemakers, who were on their way to Isfahan, and had taken up their quarters for the night in a vaulted passage which had afforded them shelter from the storm. Upon the fire they had kindled was a large caldron of savoury broth, which was boiling merrily. The long ride had given me an appetite, and I seated myself without ceremony in the group and began to help myself without waiting for an invitation. The shoemakers, although good Musulmans, made no objection to my dipping my own spoon into the mess with them. Seeing that my clothes were soaked by the rain, and that I was suffering from ague, they very civilly left me alone in the recess in which they had established themselves, and I was able to dry myself by their fire and to spread my carpet for the night by the side of its embers.

Next day we entered upon the great plain in which Isfahan is situated, and I soon came to a broad, well-beaten track, which proved the highway from Hamadan to that city. After following it for a short distance I was so exhausted by a severe attack of fever, and by the dysentery which had greatly weakened me, that I was obliged to dismount on arriving at a small village called Tunderun (?), and to take a little rest. After the shivering fit had passed I resumed my journey, but being overtaken by a heavy thunderstorm, I took refuge in a flour-mill which was fortunately hard by. The door of this building was formed by a single stone, seven feet in height and five and a half in breadth, turning upon a pivot. Such doors are common in houses and in the walls of gardens and orchards in this part of Persia. This was the largest that I had yet seen.

When the rain had ceased I again mounted

my horse, but being too unwell and weak to proceed very far, stopped for the night at Tehrun, a large village surrounded by gardens, where I was able to obtain a clean room and the repose of which I was so much in need.

The gardens amongst which I had entered before arriving at Tehrun reach in an almost uninterrupted line to Isfahan. They produce fruit and vegetables of all kinds, especially melons of exquisite flavour, which have an unrivalled reputation throughout Persia. These gardens owe their extreme productiveness to the great number of streams which descend into the plain from the Zerda-Kuh range of mountains, and are divided into innumerable rivulets for the purpose of irrigation, frequently carried underground by the tunnelled watercourses or kanâts, which have well-like openings at regular distances. These conduits are very common in Persia, and many I saw were probably of very ancient date. I passed through Najafabad, a town with avenues of fine poplars, and stopped for a short rest at the village of Seddeh.

The number of horsemen, and men and women carrying loads, whom I passed on the road showed me that I was approaching Isfahan; but nothing could be seen of the city, as it was completely buried in trees. By constantly asking my way I managed to reach, through the labyrinth of walls which enclose the gardens and melon beds, the Armenian quarter of Julfa. I had letters for M. Eugène Boré, a French gentleman, and, not knowing where to find a lodging, I presented myself to him to ask for advice. He received me with great kindness, and insisted that I should be his guest. I found residing with him M. Flandin and M. Coste, whom I had met at Taki-Bostan, and who had been attached to the special embassy which had been recently sent by the King of the French to the Shah. They had been making drawings and plans of the principal monuments, ancient and modern, in various parts of Persia. I looked forward to some pleasant and civilised society, and to that rest and care of which I had so much need after my toilsome journey, suffering severely as I was from two weakening and distressing ailments.

During my journey from Hamadan I had made careful notes of the country, taking bearings with my Kater's compass of the mountain ranges and peaks, and fixing by the same means, as well as I could, the course of streams and rivers, and the position of the towns and villages through which I passed or which I saw in the distance. I found great difficulty in obtaining the correct names of places. Whether from that inveterate habit of lying which appears to be innate in every Persian, or from suspicion of my motives in putting the question, the people whom I met on my way, and of whom I asked the name of a village, almost invariably gave me a wrong one—generally that of another in an entirely different direction. I had no little trouble in getting at the truth.

I was still suffering, when I was received by M. Boré, from a severe attack of ague and dysentery. Mr. Edward Burgess, an English merchant residing at Tabreez, who was at Isfahan on business, hearing that I had arrived, came to see me and offered to be of use to me. He proposed that we should present ourselves to the Governor, Manuchar Khan, the Mu'temedi-Dowla, or, as he was usually called, 'the Matamet,'[3] to whom he was personally known.

I was anxious to deliver the letter which had been given to me by the Haji at Hamadan for this high personage, and at the same time to lodge a complaint against Imaum Verdi Beg, my mehmandar, for his exactions and his ill-treatment of the villagers on the road. He had left me as we approached the city, taking with him the horses and donkeys laden with rice, sugar,

[3] I spell this name as it was pronounced; Mu'temedi-Dowla means 'the one upon whom the State relies.'

and other spoils which he had gathered by the use of my firman. I was determined to have him found, and, if possible, punished, and compelled to restore to their owners the animals that he had carried off, and to repay to the village chiefs the money he had levied from them.

Although very ill and weak I rode with Mr. Burgess on the second day after my arrival to the governor's palace. The Armenian suburb of Julfa is at some distance from the main portion of the city, in which only Musulmans were then permitted to live. After passing through extensive gardens we reached the Mohammedan quarters, and threading our way between mud-built houses, for the most part falling to ruins, through narrow, unpaved streets, deep in dust and mud and choked with filth and rubbish, we at length reached the Matamet's residence.

After entering, through a narrow dark passage opening into the street, a spacious yard with the usual fountains, running water, and flowers, we passed into the inner court, where the governor gave audience. The palace, which at one time must have been of great magnificence, was in a neglected and ruined condition, but had been splendidly and profusely decorated with paintings, glass, and inlaid work, such as I had seen in the palace of Douletabad. The building was thronged with miserably clad soldiers, 'ferrashes,'[4] men and women having complaints to make or petitions to present, and the usual retinue and hangers-on of a Persian nobleman in authority.

The Matamet himself sat on a chair, at a large open window, in a beautifully ornamented room at the upper end of the court. Those who had business with him, or whom he summoned, advanced with repeated bows, and then stood humbly before him as if awestruck by his presence, the sleeves of their robes, usually loose and open, closely buttoned up, and their hands joined in front—an immemorial attitude of respect in the East.[5] In the 'hauz' or pond of fresh water in the centre of the court were bundles of long switches from the pomegranate tree, soaking to be ready for use for the bastinado, which the Matamet was in the habit of administering freely and indifferently to high and low. In a corner was the pole with two loops of cord to raise the feet of the victim, who writhes on the ground and screams for mercy. This barbarous punishment was then employed in Persia for all manner of offences and crimes; the number of strokes administered varying according to the guilt

---

[4] The 'ferrash,' literally the 'sweeper,' is an attendant employed in various ways, from sweeping the rooms to administering the bastinado.

[5] The attendants of the Assyrian King are thus represented in the sculptures from Nineveh.

of the culprit. It was also constantly resorted to as a form of torture to extract confessions. The pomegranate switches, which when soaked for some time become lithe and flexible, were generally employed. The pain and injury which they inflicted were very great, and were sometimes even followed by death. Under ordinary circumstances the sufferer was unable to use his feet for some time, and frequently lost the nails of his toes. This punishment was inflicted upon men of the highest rank—governors of provinces, and even prime ministers—who had, justly or unjustly, incurred the displeasure of the Shah. I have mentioned that Hussein Khan, on his return from his special mission as ambassador to England and France, had been subjected to it on a charge of peculation.

Manuchar Khan, the Matamet, was a eunuch. He was a Georgian, born of Christian parents, and had been purchased in his childhood as a slave, had been brought up as a Musulman, and reduced to his unhappy condition. Like many of his kind, he was employed when young in the public service, and had by his remarkable abilities risen to the highest posts. He had for many years enjoyed the confidence and the favour of the Shah. Considered the best administrator in the kingdom, he had been sent to govern the great province of

Isfahan, which included within its limits the wild and lawless tribes of the Lurs and the Bakhtiyari, generally in rebellion, and the semi-independent Arab population of the plains between the Luristan Mountains and the Euphrates. He was hated and feared for his cruelty, but it was generally admitted that he ruled justly, that he protected the weak from oppression by the strong, and that where he was able to enforce his authority life and property were secure. He was known for the ingenuity with which he had invented new forms of punishment and torture to strike terror into evil-doers, and to make examples of those who dared to resist his authority or that of his master the Shah, thus justifying the reproach addressed to beings of his class, of insensibility to human suffering. One of his modes of dealing with criminals was what he termed 'planting vines.' A hole having been dug in the ground, men were thrust headlong into it and then covered with earth, their legs being allowed to protrude to represent what he facetiously called 'the vines.' I was told that he had ordered a horse-stealer to have all his teeth drawn, which were driven into the soles of his feet as if he were being shod. His head was then put into a nose-bag filled with hay, and he was thus left to die. A tower still existed near Shiraz which he had built of three hundred living men belonging to the Mamesenni,[6] a tribe inhabiting the mountains to the north of Shiraz, which had rebelled against the Shah. They were laid in layers of ten, mortar being spread between each layer, the heads of the unhappy victims being left free. Some of them were said to have been kept alive for several days by being fed by their friends, a life of torture being thus prolonged by a false compassion. At that time few nations, however barbarous, equalled—none probably exceeded—the Persian in the shocking cruelty, ingenuity, and indifference with which death or torture was inflicted.

The Matamet had the usual characteristics of the eunuch He was beardless, had a smooth colourless face, with hanging cheeks and a weak, shrill, feminine voice. He was short, stout, and flabby, and his limbs were ungainly and slow of movement. His features, which were of the Georgian type, had a wearied and listless appearance, and were without expression or animation. He was dressed in the usual Persian costume—his tunic being of the finest Cashmere—and he carried a jewel-handled curved dagger in the shawl folded round his waist. He received us courteously, said a few civil things about the English nation, which he distinguished from the English Government,

[6] A contraction of Mohammed Husseini.

and invited us to come up into the room in which he was seated and to take our places on a carpet spread near him.

I handed him my firman and the letter from the Haji, and being unable to suppress my indignation against Imaum Verdi Beg, my mehmandar, for his ill-treatment of the villagers on the road, I denounced him at once in vehement terms, describing his misconduct and the insolent manner in which he had behaved to me when I had remonstrated against it. He applied a variety of opprobrious and foul epithets to the Ghulâm himself, and to his mother and all his female relatives, after the Persian fashion, and promised that he should receive condign punishment. And he was as good as his word, for two days after Imaum Verdi came hobbling to me with a very rueful countenance, and his feet swollen from the effects of the bastinado which he had received. I was inclined to pity the poor wretch, although he had richly deserved his punishment, but I almost regretted that I had denounced him to the Matamet when he said to me in an appealing tone, 'What good, sir, has the stick that I have eaten done you? Who has profited by it? You and I might have divided the money and the supplies that, as the Shah's servant, I was entitled by his firman to obtain for you on our way. The villagers would have been none the worse, as they would have deducted the amount from their taxes. Do you think that they will get back their horses, or their donkeys, or their tomans? No, the Matamet has taken them all for himself. He is a rich man and does not want them; I am a poor man and do. He is the greater robber of the two. He goes unpunished and I have scarcely a nail left on my toes.'

After the Matamet had made the usual inquiries as to the object of my journey, and as to the route I desired to take and the places I wished to visit, he said that had I been accompanied by a competent Ghulâm I should have met with no difficulty in carrying out my original intention of crossing the Bakhtiyari Mountains to Shuster. He promised to send with me one of his own officers, who would conduct me to that city. A Bakhtiyari chief, named Shefi'a Khan, who happened to be present, confirmed what the governor had said, and informed me that one of the brothers of Mehemet Taki Khan, the great Bakhtiyari chief, was then in Isfahan. When I took my leave of the governor he told me that my new mehmandar would be ready to leave immediately, and that I should receive the letters he had promised me without delay.

The day after my interview with the Matamet I succeeded after some trouble in finding Shefi'a

Khan, who had promised to introduce me to Ali Naghi Khan, the brother of the principal chief of the Bakhtiyari tribes. They were both lodging in the upper story of a half-ruined building forming part of one of the ancient royal palaces. The entrance was crowded with their retainers—tall, handsome, but fierce-looking men, in very ragged clothes. They wore the white felt skull-cap, sometimes embroidered at the edge, when worn by a chief, with coloured wools, common to all the Lurs— their heads being closely shaven after the Persian fashion, with the exception of two locks, called 'zulf,' one falling on each side of the face. The Bakhtiyari usually twist round the skull-cap, in the form of a turban, a long piece of coarse linen of a brown colour, with stripes of black and white, called a 'lung,' one end of which is allowed to fall down the back, whilst the other forms a top-knot. In other respects the Lurs wear the usual Persian costume, but made of very coarse materials, and, as a protection against rain and cold, an outer, loose-fitting coat of felt reaching to the elbows and a little below the knees. Their shoes of cotton twist, called 'giveh,' and their stockings of coloured wools, are made by their women. A long matchlock—neither flintlocks nor percussion-caps were then known to the Persian tribes—is rarely out of their hands. Hanging to a leather belt round their waist they carry a variety of objects for loading and cleaning their guns — a kind of bottle with a long neck, made of buffalo-hide, to contain coarse gunpowder; a small curved iron flask, opening with a spring, to hold the finer gunpowder for priming; a variety of metal picks and instruments; a mould for casting bullets; pouches of embroidered leather for balls and wadding; and an iron ramrod to load the long pistol always thrust into their girdles. I have thus minutely described the Bakhtiyari dress as I adopted it when I left Isfahan, and wore it during my residence with the tribe [7]

I had some difficulty in making my way to Shefi'a Khan through this crowd of idlers, who were not a little surprised at learning that I was a Christian, and especially a 'Feringhi,' as they had never seen one before, and were evidently not quite certain as to how they should treat me. I found the Khan in a small room at the top of a rickety wooden staircase. He received me very civilly, and conducted me at once to Ali Naghi Khan. The Bakhtiyari chief was seated on a felt rug, leaning against a bolster formed of his quilt and bed-clothes rolled up in a piece of chequered silk. In front of him was a large circular metal tray, on which were little saucers containing various

[7] See the frontispiece to this volume.

kinds of sweetmeats and condiments. In one hand he held a small porcelain cup, from which he occasionally sipped 'arak,' and in the other a 'kaleôn,' from which he drew clouds of smoke. An effeminate youth was singing verses from Hafiz and other Persian poets, accompanied by a man playing upon a kind of guitar. Ali Naghi Khan had unwound the shawl from his waist, and had unbuttoned his shirt and his robe, and on his closely shaven head, jauntily stuck on one side, was a small triangular cap made of Cashmere shawl. It was evident that the chief was indulging in a debauch with four or five friends who were seated near him But he had still his senses about him. I was amused at seeing in one of the corners of the room a mulla squatted upon his hams, and rocking himself to and fro whilst reading from the Koran, and interspersing with the text loud ejaculations of 'Ya Allah!' and 'Ya Ali!' apparently unmindful of the violation of the laws of his religion by his drunken associates.

Ali Naghi Khan was the second brother of Mehemet Taki Khan,[8] who at that time exercised authority over the greater part of the Bakhtiyari Mountains. He was on his way to Tehran, to be kept as a hostage for the good conduct of the

[8] This name should properly be written 'Muhammed Taghi Khan.' I write it as generally pronounced.

chief, whose loyalty was suspected, and who had recently been in open rebellion against the Shah. Shefi'a Khan had accompanied him with an escort of retainers as far as Isfahan, whence he was now about to proceed to the capital with a few attendants. He was a short, thick-set man, of about forty years of age, not ill-looking, and with an intelligent, though somewhat false, countenance.

Shefi'a Khan, kneeling down by his side, whispered to him the object of my visit. As soon as he learnt that I was an Englishman, he begged me to sit on the felt rug by his side, bade me welcome in very cordial terms, and offered me a cup of iced Shiraz wine and sweetmeats, which I could not refuse. We soon became boon companions over the bottle. It was my object to establish friendly relations with him, as I hoped through his influence and the recommendations he might give me to his brother to enter the Bakhtiyari Mountains.

We had scarcely commenced a friendly conversation when attendants entered, bearing upon their heads trays containing various kinds of pillaus, savoury stews and other dishes. The arak, the wine, and the sweetmeats were speedily removed, and the trays having been placed on the floor, the guests gathered round them, crouching on their hams. I was invited to partake of the breakfast, which was excellent. Persian cookery is superior to that of any other Eastern nation. As I was a Frank and an infidel I had a tray to myself, an arrangement to which I by no means objected, although I could never altogether get over the sense of humiliation at being treated as unclean and unfit to dip my fingers into the same dish with true believers.

After the breakfast had been removed and the usual kaleôns smoked, the Khan spoke to me about my contemplated journey to the Bakhtiyari Mountains. He had already been at Tehran, where he had acquired the manners and the vices of the Persians who frequented the court. As he had seen Englishmen in the capital and had learnt something about their habits and customs, I was able to make him understand the object of my journey, and to remove the impression that might have existed in his mind that I was a spy, or that I was travelling in search of buried treasures, or for the discovery of a talisman which would enable the Franks to conquer his country—for such are the usual reasons assigned by wild tribes like the Bakhtiyari to the presence of Europeans amongst them. These suspicions have more than once led to fatal results. He very readily answered some questions I put to him as to various ruins of which I had heard, and when he was unable to give me the information I required, he sent for such of his attendants as might be able to supply it. He expressed regret that he was not returning to the mountains, otherwise, he said, I should have accompanied him, but he promised to give me a letter to his brother, and suggested that I should join Shefi'a Khan, who would shortly leave Isfahan for Kala Tul, the residence of Mehemet Taki Khan. I gladly availed myself of his offer. It would indeed have been impossible to have found a better opportunity for visiting this then unexplored region. I returned to Julfa well satisfied with my day's work, hoping to be able to resume my journey without delay.

The Matamet had met my request to be allowed to proceed either through Yezd or Kerman to the Seistan with so absolute a refusal, that I thought it better to renounce for the time any attempt to reach that district from Isfahan. The news of the occupation of Afghanistan by the British troops had caused great excitement in Central Asia, and had added greatly to the insecurity of the country on the eastern borders of Persia. It was reported that the people of Herat, backed by the English, had taken Ghurian; that Kerman had fallen into the hands of the Belooches, who had murdered the governor; and that Aga Khan, a descendant of the 'veiled prophet,' much venerated in Southern

Persia, was marching with a considerable force against the Shah, supported by the British Government. This man was a restless intriguer, who had for a long time kept the country in a state of disquiet. He had been made prisoner some time before when in rebellion to the Shah, but had been released on account of his sacred character at the intercession of the Haji. He was now again in arms and was said to be advancing upon Yezd and Kerman. The death of Dr. Forbes, who had been murdered in an attempt to reach the Lake of Furrah, was known to the Matamet, and he was persuaded that I should meet with the same fate, and that he would be held responsible for anything that might happen to me if he permitted me to undertake so dangerous an expedition. As he had the means of preventing me from carrying out my intention of going to Yezd, I decided upon waiting until the state of affairs, which at that time was unquestionably very unsettled, might enable me to persevere in it. In the meanwhile I could employ my time usefully in exploring the Bakhtiyari Mountains, and in endeavouring to solve some interesting geographical and archæological problems. Such were the reasons which induced me to renounce for the time my original plan of reaching Kandahar through the Seistan, but I was still resolved to adhere to it unless I found insurmountable difficulties in my way. Alone, and without official or other protection—England being in a state of war with Persia—and suspected of being a spy or an English agent, I was under the necessity of acting with extreme prudence and caution, although I was prepared to run any risk that the object I had in view should appear to me to justify.

Although Shefi'a Khan had assured me that he was about to leave Isfahan at once, the days passed by without any signs of his departure. I was continually going to and fro to the caravanserai, a ruined building in the middle of the city, to which he had removed after the departure of Ali Naghi Khan. There he sat, imperturbably smoking his kaleôn on a raised platform of brickwork in the centre of a dirty yard in which his horses and mules and those of other travellers were tethered, and in which the smells were consequently almost intolerable. He had always some excuse ready to explain the delay in his departure. At one time it was a hostile tribe that had closed the road; at another, he was endeavouring to raise, by the sale of his effects, money to pay his bill at the khan and to provide for the necessary expenses of his journey. Then the mulla who was to accompany him had failed, after opening the Koran and other books and consulting the first words on the page,[9] to name the day on which it would be propitious to begin the journey. His detention was, however, mainly caused by discussions with the Matamet, who wished to send one of his officers to collect the tribute from the Bakhtiyari tribes, and who was disposed to retain the Khan as hostage for its payment.

As my patience was almost exhausted by these constant delays, I applied to the Matamet to allow me to proceed to the mountains accompanied by one of his Ghulâms, as he had proposed at my first instance. But he now alleged that the condition of the Bakhtiyari tribes was such that I could not travel amongst them, except in company with Shefi'a Khan, and that if I was still resolved upon visiting Shuster I must wait until he was ready to leave. There was nothing to be done but to resign myself to this detention, which lasted

[9] This mode of ascertaining the propitious moment for commencing an undertaking or a journey prevails among Mohammedans, and is called 'Istikâra.' The Koran, or a volume of verses of Hafiz or Saadi or some other poet, is opened at random, and the first words or sentences which occur at the top of the right-hand page are supposed to decide the question, their sense being generally interpreted and applied to the occasion by a mulla, although any one may do so for himself. Another mode is that of separating at hazard, and with closed eyes, a number of beads from the rest in the chaplet which Persians are in the habit of carrying. If the number thus separated be odd it is considered unfavourable; if even, the contrary.

for nearly five weeks. Persians, like other Orientals, have no conception of the value of time, and my Bakhtiyari friends could not understand my impatience to get away from a place which, according to their ideas, afforded so many sources of delight and enjoyment.

My time at Isfahan was not idly or unpleasantly spent. I continued to study the Persian language. In the house of M. Boré I found agreeable and instructive society. He was himself an accomplished man, well versed in Oriental languages. He was young, of independent means, and a religious enthusiast very ardent and zealous in promoting the political interests of his country—a kind of politico-religious propagandism much encouraged in the East by all French governments. His object in establishing himself at Isfahan was to open schools in the Armenian quarter of Julfa for making converts to Roman Catholicism. He had not, I believe, achieved much success. A Roman Catholic mission had been established for many years in this Christian suburb of Isfahan. It was chiefly directed by Italian priests connected with the Propaganda at Rome. M. Boré worked with them, and I frequently met them at his house. They were ignorant, narrow-minded men, but jovial companions, made excellent wine from the grapes of the country, and liked good cheer. One of them had been employed in other parts of Persia, principally at Urumiah, in endeavouring to convert the Nestorians. He spoke of these interesting Christians with great contempt. His plan for turning them to the true faith, which he would describe with a knowing twinkle of the eye, might, if steadily carried out, have proved effective. 'You are aware,' said he, 'that the Nestorian tribes which inhabit the high mountains of Luristan are under the necessity of descending into the plains in winter, otherwise they and their flocks would starve. I shall suggest to the Persian authorities to prevent them from doing so, except on the condition that they abandon their heretical creed and become good Catholics. They must then either perish or be converted, and they will no doubt adopt the latter alternative.' I ventured to observe that although this method of conversion might prove, as he expected, successful, it was not altogether in accordance with European ideas, but savoured somewhat of Persian modes of proceeding. He replied, with simple earnestness, 'Bah! caro mio, con queste bestie non si fanno tante ceremonie.' Such heroic modes of conversion were not uncommon among those who were employed in propagating the Roman Catholic faith, and French political interests at the same time, in the East.

M. Boré, with all his learning and enlightenment,

was a religious fanatic and profoundly intolerant of heretics. After residing with him for a fortnight, and having been treated by him with great kindness and hospitality, I found myself compelled, to my great sorrow, to leave his house under the following circumstances. The Embassy which the King of the French had sent to the Shah had not succeeded in obtaining the object of its mission, and had left Persia much irritated at its failure, which was mainly attributed by it and the French Government to English intrigues. The truth was, I believe, that they had been duped by Hussein Khan, who had been sent as ambassador to Paris. The subject was an unpleasant one for me to discuss, and I avoided it in conversation with my host. One day, however, at dinner, it was raised by M. Flandin, the French artist, who denounced my country and countrymen in very offensive terms, M. Boré himself joining in the abuse. They accused the English Government and English agents of having had recourse to poison to prevent Frenchmen from establishing themselves and gaining influence in Persia, and of having actually engaged assassins to murder M. Outray, when on his way on a diplomatic mission to Tehran I denied, with indignation, these ridiculous and calumnious charges, and high words having ensued, I moved from M. Boré's house to a ruined building occupied by Mr. Burgess. I afterwards met M. Boré at Constantinople. He had then formally joined the Society of Jesus, and was at the head of a Jesuit establishment in Galata.[1]

M. Coste, who, like M. Flandin, had accompanied the French Embassy, to prepare the great work which was to be published at the expense of the French Government, to illustrate the monuments and antiquities of Persia, was of a very different disposition from his vain, impetuous, and aggressive companion. He was an architect from Marseilles, and had already gained a reputation by a remarkable book on the Arab architecture of Egypt. He was an accurate and skilful draughtsman, and a man of simple and amiable character and habits, so thoroughly absorbed in his work that, when engaged in it, he was too absent to think of anything else. On one occasion I suggested to him to make a drawing of a finely carved capital of the Sassanian period, which I had discovered in one of my wanderings in an out-of-the-

[1] The alleged forced conversion by him of an Armenian caused a tumult in Julfa which nearly led to his assassination, and he was compelled to leave Isfahan. When I found myself in serious difficulties some months afterwards at Shuster, where I was a virtual prisoner in the Matamet's hands and without money, I contrived to send a letter to M. Boré describing my position. I received from him a reply which proved that, although a bigot in religion, he was a man of a kindly and generous disposition.

way part of the city. He rode off at once to do so. Dismounting, he seated himself on a stone, and passing his arm through his horse's reins, commenced his sketch. After finishing it he found to his surprise that his horse had disappeared. A thief had slipped the bridle off the animal's head, and had led it away, leaving the reins on the artist's arm. I went with him to complain of the theft to the Matamet, who burst into a fit of laughter when he heard the story. 'That must have been the work of a Bakhtiyari,' he exclaimed, 'the most skilful and audacious of thieves No one else could have imagined or executed such a trick.' He sent off at once to Shefi'a Khan, threatening to bastinado his followers all round unless the horse was restored within a few hours. He was right in his conjecture. One of the chief's attendants had been the thief, and the animal was duly returned to its owner.

During my residence at Isfahan I passed much of my time visiting the mosques (into which, however, I could not, as a Christian, enter), and the principal buildings and monuments of this former capital of the Persian kingdom, which had been deserted by the court for Tehran. I was delighted with the beauty of some of these mosques, with their domes and walls covered with tiles, enamelled with the most elegant arabesques in the most brilliant colours, and their ample courts with fountains of cool water and splendid trees. I was equally astonished at the magnificence of the palaces of Shah Abbas and other Persian kings, with their spacious gardens, their stately avenues, and their fountains and artificial streams of running water, then deserted and fast falling to ruins. It was not difficult to picture to oneself what they must once have been. Wall-pictures representing the deeds of Rustem and other heroes of the 'Shah-Nameh,' events from Persian history, incidents of the chase and scenes of carouse and revelry, with musicians and dancing boys and girls, were still to be seen in the deserted rooms and corridors, the ceilings of which were profusely decorated with elegant arabesques. In the halls, the pavements, the panelling of the walls, and the fountains were of rare marbles inlaid with mosaic. The rills which irrigated the gardens and avenues were led through conduits of the same materials. Even the great carpets, the finest and most precious which had issued from Persian looms in the sixteenth and seventeenth centuries—unequalled for the beauty and variety of their designs and the fineness of their texture—were still spread upon the floors. The neglected pleasure-grounds were choked with rose-bushes in full bloom. Although these gorgeous ruins were desolate and deserted, they afforded the most striking proof of the luxury and splendour of the Persian court in former times. I used frequently to spend the day in wandering about them, lost in admiration at their wonderful beauty.

Mr. Burgess had several acquaintances amongst the notables of the city, who invited us to their houses and hospitably entertained us with breakfasts and dinners, at which I became acquainted with a great variety of excellent Persian dishes. They were, unfortunately, almost always accompanied by a free use of wine or arak, which generally preceded such feasts — Easterns rarely drink them during and after a repast—unless the host was a rigorous Musulman who looked upon all that intoxicates or even exhilarates as forbidden. Music and dancing were rarely wanting. The odes of Hafiz and Saadi, which have almost the same effect upon a Persian as the wine of Shiraz, were sung by professional reciters, and occasionally by some one of the company—for most educated Persians have a rich store of them in their memory.

But the most characteristic and curious scenes of Persian life were those I witnessed in the house of a Lur chief who had left his native mountains and had established himself in Isfahan, professing to be a 'sufi,' or free-thinker. He was an intimate friend and a distant connection of Shefi'a Khan, by whom I was introduced to him. He invited me more than once to dinner, and I was present at some of those orgies in which Persians of his class were too apt to indulge. On these occasions he would take his guests into the 'enderun,' or women's apartments, in which he was safe from intrusion and less liable to cause public scandal. They were served liberally with arak and sweetmeats, whilst dancing girls performed before them. Many of these girls were strikingly handsome—some were celebrated for their beauty. Their costume consisted of loose silk jackets of some gay colour, entirely open in front so as to show the naked figure to the waist; ample silk 'shalwars,' or trousers, so full that they could scarcely be distinguished from petticoats, and embroidered skullcaps. Long braided tresses descended to their heels, and they had the usual 'zulfs,' or ringlets, on both sides of their faces. The soles of their feet, the palms of their hands, and their finger- and toenails, were stained dark red, or rather brown, with henna. Their eyebrows were coloured black, and made to meet; their eyes, which were generally large and dark, were rendered more brilliant and expressive by the use of 'kohl.'[2] Their movements were not wanting in grace; their postures, however, were frequently extravagant, and more like gymnastic exercises than dancing. Bending them-

---

[2] A black powder used to darken the eyelids.

selves backwards they would almost bring their heads and their heels together. Such dances are commonly represented in Persian paintings, which have now become well known out of Persia. These contortions soon degenerated into outrageous indecency, for these dancing girls did not refuse the wine and arak that were liberally offered to them. The musicians were women who played on guitars and dulcimers. These orgies usually ended by the guests getting very drunk, and falling asleep on the carpets, where they remained until sufficiently sober to return to their homes in the morning.

I called once or twice on the 'mujtehed,' or head mulla of the great mosque and of the Musulman religion in Isfahan. Although a very strict Mohammedan, and unwilling to be seen seated on the same carpet with a Christian—any manner of contact with an infidel rendering a follower of Islam unclean—he received me very courteously, and appeared to take pleasure in conversing with me about European manners and discoveries, and upon general politics. I always carefully avoided the discussion of subjects connected with religion—and especially controversial matters—in conversing with him and any other Persian Musulman, as an unguarded expression might have brought me into very serious trouble.

In those days the fanatical Persians were apt to deal very summarily with any one who might have used words which could be construed into an insult to their religion, or as blaspheming their Prophet. A Christian thus offending would have caused a public tumult, and might even have been torn to pieces.

CHAPTER VIII.

Departure from Isfahan—My travelling companions—The Shutur-bashi—Shefi'a Khan — False alarm — Enter the Bakhtiyari country—Fellaut—Hospitable reception—Chilaga—A foray—Lurdagon—A Bakhtiyari feast—Effect of poetry—Difficult mountain pass—Thieves—Reach the Karun—Kala Tul—The guest-room—Mehemet Taki Khan's brothers—His wife—His sick son—The great Bakhtiyari chief—Cure his son—Khatun-jan Khanum—Khanumi—Fatima—Hussein Kuli—Ali Nagi Khan's wives—Dress of Bakhtiyari women—Marriages—Life at Kala Tul—The Bakhtiyari.

ON September 22 Shefi'a Khan sent to tell me that everything was ready for his departure, that a mulla of recognised sanctity had declared, after consulting the Koran and his beads, that the day was propitious for undertaking the journey, and that he intended to leave Isfahan that very evening for the mountains. It was his intention, he said, to travel by night, as the heat was still great, and as it would be safer to do so to avoid the marauders, who were believed to infest the country through which we had to pass. He proposed that I should meet him at sunset in the garden of the ruined palace of Heft-Dest, near the

Shiraz bridge. I was there by the time appointed Instead of finding, as I had expected, the Khan and his companions ready to start, I saw that they had evidently settled themselves down for the night. The chief, with his eternal kaleôn, was seated on his carpet under a tree; the women who were to accompany us were crouched amongst the baggage, enveloped from head to foot in their thick 'chaders,' or veils; the horses and mules were tethered for the night, and the men were occupied in giving them their provender. Shefi'a Khan apologised for the delay, throwing the blame for it upon an officer of the Matamet, who was to accompany us to the mountains upon some business connected with the revenue, and who had sent word to say that he could not join us until the morning. He should have, therefore, to give up his intention of travelling by night in order not to lose another day, and whether the 'shutur-bashi,' chief of the running footmen—for such was the title of the official for whom we were waiting —appeared or not, he was determined to start at dawn.

There was nothing to be done but to picket my horse, and to spread my carpet as near to it as possible, so as to be on the watch for thieves. The scene was singularly picturesque. The stars were shining brilliantly overhead, the majestic trees of a long avenue rose darkly above us, a bright fire threw a red and flickering glare upon the countenances of the wild and savage men gathered above it, and the silence was soon only disturbed by the tinkling of the bells of the mules tethered about us. I wrapped myself in my cloak, for the nights were beginning to be cold, and soon fell asleep.

I was awoke before daylight by the noise and bustle of the preparations for our departure. The shutur-bashi had arrived. The attendants were placing the baggage on the mules, and the women and children on the top of the loads. I saddled my horse, and mounting, joined the small caravan. It was a motley company. Shefi'a Khan, who belonged to the Suhunni, a division of the great Bakhtiyari tribe of Chehar Lang, was handsome, tall, and of a commanding presence. He wore the Lur dress, except that since his visit to Isfahan he had laid aside the felt skull-cap for the 'kulâh,' or tall lambskin Persian hat, as more becoming and dignified. It would be difficult to imagine a more wild and ferocious-looking set of fellows than his followers; but they were very fine specimens of the human race, like most of the mountain tribesmen of Persia, who claim to be of pure Persian or Aryan blood, and to descend from the ancient inhabitants of the country they still occupy. Two ladies, wives of Ali Naghi Khan, the brother of the great chief, who would not accompany their husband to Tehran, were returning with us to their home with their maids. They rode on mules, perched high up on the baggage. At first they were closely veiled, their faces being concealed by a kind of network. But they soon dropped their veils and their reserve, and we became friends, talking on the way and when resting for the night. They were both really beautiful women. One of them had a little daughter, a lovely child about five years old, with large black eyes and long silken eyelashes. She and I became fast allies. She would make me take her on my saddle as we rode along, entertaining me with her merry chatter, and when we rested she would sit on my carpet and play with my watch or compass. She was adorned with all the trinkets that her mother had been able to save from the pawnbrokers of Isfahan, her little feet and hands were dyed with henna, and her wrists and ankles encircled with numerous gold and silver bangles. Her name was Bibi Mah —Lady Moon.

The shutur-bashi was one of those vain, lying, and unprincipled fellows who abounded in Persia and were plentifully found in the public service. He was a mirza, or scribe, rode a strong, sturdy mule, smoked, through a long flexible leather tube, a fine enamelled kaleôn which was carried by an attendant on horseback at his side, and had half a dozen servants. He gave himself great airs, and seemed to avoid the rest of the company.

I had adopted, as I have mentioned, the Bakhtiyari costume with the outer coat of felt. Shefi'a Khan suggested that I should do so to avoid being recognised in the dangerous country through which we had to pass, where a European had never before been seen. He begged me to have my gun and pistols always ready, as they were meant not for show but for use, and as we might have occasion for them at any moment during our journey. I had succeeded in obtaining from a banker at Isfahan a small sum, about twenty tomans, or 10$l.$, in gold, which I carried, as usual, in a belt worn round my waist next to the skin. This was all the money I possessed. My Bakhtiyari friends had insisted that I should require none when with their tribe, with whom hospitality was a duty, and who would resent as an insult an offer of payment for it. I was advised, indeed, in order not to tempt the cupidity of the evilly-disposed, to take no money with me, and I thought it advisable to conceal what little I had. Shefi'a Khan had given me as my personal attendant—who was to be responsible for my person and goods—a youth with stern and fierce features, named Khunkiar.[1] All my effects were contained in a pair of saddle-bags, worked in worsted of divers colours, which I could place on my Persian saddle. They only consisted of a second shirt, a hammer and nails to shoe my horse, one or two books and maps, and a few necessary medicines. Shefi'a Khan allowed me to place my small carpet and wadded coverlet upon one of his baggage mules.

The caravan, which consisted of about fifty persons mostly on foot, having been formed, we commenced our march. Our progress was necessarily slow, as much of the baggage was carried by donkeys, which required continual urging with blows to make them keep up with the horses. The abuse with which a Persian assails his ass, whilst pricking the wretched animal with his dagger or a packing-needle, is indescribably foul, and could not bear literal translation. It is usually applied to the animal's master, that is, to the driver himself. 'May your owner eat filth! May his mother, or his sister, or his father be subjected to the worst of outrages! May his grave be polluted!'

---

[1] This, as it is well known, is one of the ancient titles of the Sultans of Turkey, and is generally supposed to mean the 'blood-drinker,' and to denote their ferocious propensities; but it is, I believe, derived from a Tatar word which signifies emperor.

&c., &c.—each ejaculation being accompanied by a prod which brings blood and makes the poor beast plunge forward, generally at the same time throwing off its load. This leads to confusion and delay and to a fresh volley of abuse and of blows. The Persians are cruel to animals—in this respect differing from the Turks, who are in general kind and humane to them.

After leaving the gardens of Isfahan, which do not extend as far to the south of the city as they do to the north, we found ourselves on a broad well-beaten track, leading over a treeless, barren, and rocky country, into which extended spurs from the neighbouring hills. This was the high road leading to Shiraz. Shefi'a Khan explained to me that he had been compelled to renounce his original intention of striking into the mountains to the west of the city, and of reaching the residence of the Bakhtiyari chief by the more direct route over the range of Zerdâ-Kuh. He had, he said, received information that a tribe with which he had a blood feud, hearing that he was about to pass that way, had sent out a body of horsemen to intercept him. He had determined, therefore, to follow the main road to Shiraz as far as Koomeshah before entering the Bakhtiyari country. He would thus, he hoped, avoid all risk of meeting his enemies, who lived further to the north and who would not have been made acquainted with his change of route.

As I rode along I could abandon myself to my reflections, which were of a very mixed kind. I was much elated by the prospect of being able to visit a country hitherto unexplored by Europeans, and in which I had been led to suppose I should find important ancient monuments and inscriptions. It would have been impossible to have undertaken the journey under better auspices. I was in company of a chief of one of the Bakhtiyari tribes, who had undertaken to conduct me in safety to the residence of Mehemet Taki Khan, then their acknowledged head. Shefi'a Khan seemed well disposed towards me. I had every reason to believe that during our intercourse at Isfahan I had gained his friendship, by various little services which I was able to render him. As he had served for a short time in a regiment of regular troops organised by English officers in the Persian service, and had thus acquired some knowledge of Europeans, he did not look upon them, as ignorant Persians did in those days, as altogether unclean animals, with whom no intercourse was permitted to good Musulmans. His wild and lawless followers seemed to be disposed to be kind and friendly to me, and I had no cause to mistrust them. But the Bakhtiyari bear the very worst reputation in

Persia. They are denounced as a race of robbers — treacherous, cruel, and bloodthirsty. Their very name is held in fear and detestation by the timid inhabitants of the districts which are exposed to their depredations. I had been repeatedly warned that I ran the greatest peril in placing myself in their hands, and that although I might possibly succeed in entering their mountains, the chances of getting out of them again were but few. However, I was of good cheer— very hopeful and very confident that my good fortune would not desert me, and that by tact and prudence I should succeed in coming safely out of my adventure. I determined at the same time to conform in all things to the manners, habits, and customs of the people with whom I was about to mix, to avoid offending their religious sentiments and prejudices, and to be especially careful not to do anything which might give them reason to suspect that I was a spy, or had any other object in visiting their country than that of gratifying my curiosity and of exploring ancient remains. Accordingly I abstained from making notes or taking observations with my compass except when I could do so unobserved. Whilst associating with my companions on intimate terms, and conversing freely with them, I abstained from touching their food and their drinking vessels unless invited to do so, and from showing too much curiosity and asking too many questions about their country, its resources, and the roads through it. Shefi'a Khan himself was more enlightened and liberal in his opinions than other Bakhtiyari chiefs. His views had been enlarged by his visits to Tehran and Isfahan. He was always ready to give me such information as I required, and did not appear at any time to suspect my motives in asking for it. He could even understand a map and the use of a compass, and I could explain to him the object of my researches. He could read and write—rare accomplishments amongst his fellow-tribesmen—and knew by heart a considerable part of the 'Shah-Nameh,' and the odes of the great Persian poets. He took pleasure in reciting verses from them to me as we rode together. I could not have been introduced to the great Bakhtiyari Khan under better auspices or more favourable circumstances; for neither my firman nor my letter from the Matamet would have availed me anything with one of the most powerful chiefs in Persia, who boasted that he owed no allegiance to the Shah.

We stopped for the first night in a ruined caravanserai in the village of Mayar. Shefi'a Khan's purse had been entirely drained by his long detention at Isfahan, and he had sold or pawned all the little property that he had taken there with him. He was consequently unable to buy provisions, and was compelled, as were his followers, to be content with the dry bread that they had brought with them. As I did not wish to make any display of my money, or to do otherwise than my companions, I had to be satisfied with the same fare, to which, however, I added some delicious grapes. The shutur-bashi being a Government officer, and little inclined to go to bed on so frugal a meal, quartered himself upon the principal inhabitant of the place, and by threats, backed up with the whip, obtained a good supper.

We fared no better the next evening at the small town of Koomeshah. On the following day we had scarcely ridden for an hour when we halted at the village of Babakhan. As the shutur-bashi had presented Shefi'a Khan with a sheep which he had exacted from the villagers, it was decided that we should proceed no farther, but stop and make a feast.

We had now left the high road to Shiraz, and were approaching the mountains. Our next sleeping-place was Coree, a small village surrounded by a mud wall, within which we had to take up our quarters, as the place was exposed to attack from Bakhtiyari marauders. About midnight we were roused by an alarm that horsemen had been seen in the distance. Shefi'a Khan and his followers armed themselves, and after putting the women into a stable for safety, sallied forth to meet the enemy. There was a good deal of noise and much firing of guns; but after a time the horsemen, if there had been any, withdrew. We returned to our carpets, and passed the night without being again disturbed.

However, we learnt from the villagers that a party of Bakhtiyari belonging to a tribe hostile to that of Shefi'a Khan was in the neighbourhood, and that there was reason to apprehend that our caravan might be attacked when crossing the difficult and precipitous mountain range which separated us from the plain of Semiroon. Consequently, in order to be prepared, our caravan on starting was formed into a line of march in which it was to continue during the day. The ketkhudâ, or headman, of Coree, furnished us with an escort of horsemen, and we were joined by a number of men with laden donkeys, who had been waiting for an opportunity to pass through this dangerous country in company with other travellers. We now mustered about five-and-thirty well-armed horsemen, and about twenty matchlock-men on foot. The shutur-bashi rode in front with a part of this force; the women and children, with the baggage and the laden donkeys, guarded by matchlock-men, formed the centre; and Shefi'a Khan, with his Bakhtiyari followers, brought up the rear, which was considered the post of danger. Horsemen were sent out as scouts to watch the movements of the enemy. Our pursuers appeared once or twice at a distance, and shots were exchanged with them, but they did not venture to attack us. It was fortunate that such was the case, as the track over the mountain was narrow, rocky, and dangerous, and owing to the donkeys the confusion which prevailed was so great that we must have been at their mercy. I was well pleased when we had reached the crest of the pass, and commenced a very precipitous descent to the village of Semiroon, which lay beneath us. To the south of the plain of Semiroon, the mountains which form the principal seat of the Bakhtiyari tribes rose in grand snow-covered peaks. We issued from a narrow gorge wooded with magnificent walnut trees, and arrived in the afternoon at the village, which had once been a town of some importance, inhabited by about a thousand families, reduced at that time to three hundred. Its houses of solid stone masonry were for the most part in ruins. We were again lodged, with the horses and mules, in the dirty courtyard of a deserted caravanserai, and had nothing but coarse dry bread for supper—our only meal during the day. My companions were longing to reach the Bakhtiyari country, where we were promised a hospitable reception and better fare.

We crossed next day a second precipitous mountain by so steep and difficult a track that we had to dismount and to walk the greater part of the way. Descending into a small plain we reached Fellaut, a Bakhtiyari village of stone huts, built at the foot of a lofty perpendicular rock. It belonged to the Duraki, a subdivision of the great Bakhtiyari tribe of Haft Lang, who only, however, inhabited it in the winter months. During the summer they migrated to the mountains, with their flocks, in search of pasture. They had descended from the high lands on the approach of autumn, and were encamped in the plain, over which were scattered their black tents. The sight of them gave infinite delight to Shefi'a Khan and his followers, who looked forward to finding among the 'Iliyât,' or livers in tents, that hospitality which was not extended to travellers by the inhabitants of the towns and villages of the plains. They were not disappointed. As soon as our approach was announced the chief of the encampment came out to meet us, and invited us to spread our carpets near a stream, under some fine trees. As the sun went down trays were brought from his tent with excellent pillaus, which were very welcome after our long fast. Shefi'a Khan sat to a late hour in the night, surrounded by the principal men of the tribe, inquiring about the various events which had occurred during his absence—the 'chapous,' or forays, the tribal feuds, and the death of friends in war or by the assassin's hand— and discussing the affairs of the mountains in general. The Duraki were at that time at peace with his tribe, the Suhunni, which is a branch of the other great division of the Bakhtiyaris known as the Chehar Lang. They had frequently been deadly enemies. Although all the sub-divisions of the Bakhtiyari clan which occupy this part of the mountains then acknowledged the supremacy of Mehemet Taki Khan, they were constantly engaged in bloody quarrels arising out of questions of right of pasture and other such matters. When they were thus at war they ruthlessly pillaged and murdered each other. With them 'the life of a man was as the life of a sheep,' as the Persians say, and they would slay the one with as much unconcern as the other. Had there not been peace at that moment between the Suhunni and Duraki, Shefi'a Khan would not have ventured into their tents.

The mountainous country beyond Fellaut, in which we now entered, was thickly wooded with the 'beloot,' or oak. I observed several different species, one in particular bearing a very large and handsome acorn. But these trees are principally valuable for the white substance called by the Bakhtiyari 'gaz,' or 'gazu,' a kind of manna, which is deposited, I believe, by an insect, upon the leaves of the tree. It is an article of export to all parts of Persia, and is everywhere sold in the bazars, and employed in the manufacture of a sweetmeat called 'gazenjubin,' which is much relished and considered very wholesome. When boiled with the leaves and allowed to harden, it forms a kind of greenish cake not disagreeable to the taste But prepared for the use of the ladies of the enderun, and to be offered to guests, it is carefully skimmed and separated from the leaves, when it becomes a sort of white paste of very delicate flavour.

In the valley of Chilaga, which we entered about midday, there were vineyards and corn-fields and some good arable land. But we found the village of that name in great commotion—men and women were gathered together in knots, gesticulating violently, and screaming at the top of their voices. A hostile tribe had that morning driven off part of their flocks and herds. Horsemen had galloped in all directions to pursue the marauders and to endeavour to recapture their booty. There was great excitement when they returned, bringing back some of the cattle that had been stolen, and a prisoner, who was handed over to Shefi'a Khan to be taken to the castle of a neighbouring chief. After breakfasting upon luscious grapes and bowls of honey we continued our journey, the captive being driven before us with his arms bound behind his back. His mother —an aged woman, with long dishevelled grey locks—followed us with loud cries, supplicating for his release, and appealing to the Bakhtiyari ladies on his behalf. When she found that her entreaties were of no avail, she broke out in maledictions and curses upon our heads, and, throwing herself upon her son, endeavoured to unloose his arms until she was driven back.

On our way we saw at a distance some oxen and donkeys grazing. They were recognised as belonging to the freebooters who had that morning plundered the inhabitants of Chilaga. Shefi'a Khan consequently considered them as lawful prey, and dashing with some of his horsemen through a rapid stream which separated us from them, he drove them over it. With this addition to our caravan the confusion of our march was much increased in a narrow gorge, through which we had to pass as fast as we could, anticipating that we should be attacked. We heard the sound of matchlock-firing in our rear, and passed horsemen hurrying to the fray, amongst them five young men of savage appearance, belonging to the castle of Lurdagon, which we reached in the afternoon. Its owner, one Ali-Gedâ Khan, a Bakhtiyari chief, garishly dressed, and followed by a number of well-armed retainers, came out to meet us, and warmly welcomed Shefi'a Khan. We sat down with him under some spreading trees, on the bank of a stream. Such cool and shady spots are generally found near Persian villages for the resort of the inhabitants and of travellers during the heat of the day. We were surrounded by lofty rugged mountains, and the castle with its towers, like a feudal stronghold of the Middle Ages, stood on the outskirts of a gloomy forest. It was altogether a very picturesque and romantic spot, rendered even more so by the crowd of ferocious and savage-looking men, all armed to the teeth, who gathered round us. Our host's reputation for hospitality, of which I had heard much on the way, was not belied. Two hours after sunset a procession of attendants, carrying torches, issued from the gate of the castle bearing trays on their heads, with an excellent and ample supper of pillaus, boiled and roasted meat, fowls, melons, grapes, sherbets, curds, and other delicacies, which did honour to the enderun of the chief whose ladies had prepared our repast. I entertain a lively recollection of the feast and of the scene; for it was the first time that I had been the guest of one of the principal mountain chiefs, and his appearance, his independent and manly bearing, and the quiet dignity of his manners, so different from those of the false and obsequious Persians of the towns, much impressed me. The prisoner was locked up in the castle, but, owing to the negligence or connivance of his guards, effected his escape during the night.

We remained at Lurdagon during the following day to rest our horses and baggage animals, which had been much tried by the steep, stony mountain passes which we had crossed. It soon became known that I was a Frank, and as all Franks were believed to be cunning physicians, I was visited by men and women asking for medicine for various complaints, chiefly intermittent fever, which is very prevalent in the valleys during the autumn. I was invited to visit the ladies of the chief, who received me in the enderun without being veiled, and asked me to prescribe for them and their children—charms for securing the affections of their husbands and to enable them to bear children being principally in request. They brought me trays of fruit and sweetmeats, and similar attention was shown me even in the humblest tents that I entered. Although I was the first European who had visited these wild mountains, and none of the people of Lurdagon had ever seen a Christian before, I was everywhere treated with kindness and respect, although I was naturally an object of curiosity. It was with some difficulty that I was able to leave the encampment unfollowed, to indulge in a bath in the cool and refreshing stream. I was much in need of one, as my clothes had not been off my back since I had left Isfahan.

The castle of Lurdagon stands on an artificial eminence in an island formed by a stream, which is one of the sources of the Karun, here called Bugûr. I was told by Shefi'a Khan, who was well informed as to the history of his native mountains, that it occupied the site of an ancient town which was, at one time, the capital of the extensive district of Luri Buzurg (the greater country of the Lurs), now inhabited by a variety of tribal subdivisions, each having its own chief, but all acknowledging the supremacy of Mehemet Taki Khan. Near the village there was a large artificial mound which may cover some ancient ruins. The castle, which was square, consisted of five circular bastions, united by a curtain. Within was the house in which the family of Ali-Gedâ Khan resided. The arched gateway and the tops of the towers were adorned with various trophies of the chase, such as the skulls and horns of the ibex, or wild mountain goat, and the antlers of a large stag. The place may have been sufficiently strong to resist the attacks of the Bakhtiyari, but not those of a regular force. The chief, whose jurisdiction was extensive, could assemble for its defence and for purposes of war a considerable body of horsemen and of matchlock-men.

In the evening Ali-Gedâ Khan produced a splendidly illustrated manuscript of the Nizami, a present to his father, a well-known Bakhtiyari chief, from Feth-Ali Shah. Shefi'a Khan read portions of it aloud, and recited verses describing the loves of Khosrau and Shirin, to an admiring and excited group of wild-looking men who stood in a circle round him, leaning on their long matchlocks. They followed him with intense interest, expressing their sympathy for the lovers with deep-drawn sighs, and their admiration for the heroic deeds of Khosrau with violent gesticulations and cries of approval. I often afterwards witnessed the effect thus produced by the recital of poetry upon these savage but impressionable mountaineers. The scene, lighted up by the bright fire round which we were sitting, in the midst of forests and mountains towering into the sky above us, was a very strange and striking one.

We left Lurdagon accompanied by our hospitable

host, who had entertained us magnificently, considering that he was only the chief of a small tribe, with no other resources except what his mountains could supply. He was followed by some fifty well-armed and well-mounted retainers. He never stirred without them, having blood-feuds with most of his neighbours, whose relations he had killed in war, or perhaps put out of the way in his struggle for the chieftainship. We had a delightful ride of four hours along the banks of the Bugûr, here a considerable stream, winding through the valley and forcing its way through narrow and precipitous gorges. I was surprised to find the valley so highly cultivated and so fertile. The stream was used to irrigate melon-beds and extensive rice-fields. On the lower slopes of the mountains grew corn and barley, and fruit and other trees abounded. Higher up were dense forests of oak. Black tents and small hamlets for winter habitation were scattered on all sides. Horsemen came from them to accompany us on a part of our road.

We arrived early at the encampment of Hussein Aga, the brother of Ali-Gedâ Khan. He was living with his wives and children, and his immediate followers, in black tents and huts formed of boughs, in a pleasant glade in the forest. He had not yet occupied his winter quarters—a small village with a castle hard by. He received us with the same hospitality that we had experienced from his brother.

We had still to cross the highest mountains which divided us from the valleys and plains of Khuzistan, or Susiana. The ascent commenced immediately after we left the encampment in which we had passed the night. We followed a track which was more fitted for the mountain goat than for either horses or mules. The poor beasts were continually stumbling and falling, and the way was impeded by their loads, which had to be readjusted or replaced. We had to lead them with great toil over loose stones or along the face of precipitous rocks as smooth and slippery as glass. Once my horse, losing its footing, rolled down a steep slope for about thirty feet. It was fortunately stopped by some bushes before it had reached the edge of a precipice, and was dragged back by main force by Shefi'a Khan's men, having fortunately escaped with only a few cuts and bruises. The animals suffered greatly, and our path was marked by their blood. The women, wrapped in their veils and wearing their clumsy travelling dress, could scarcely keep up with the rest of the company, and were overcome with fatigue. The road which Shefi'a Khan had taken was, I believe, rarely if ever followed by caravans, and could scarcely be called practicable for horsemen. It was chosen by him in order to avoid a district inhabited by a tribe with which he was at enmity.

From the summit of the pass, which we reached with the greatest labour and difficulty, and where we stopped to rest, we had a magnificent view of various lofty mountain ranges and snow-capped peaks. Descending by a track less difficult and dangerous than the one we had followed on the ascent, we reached a narrow marshy valley, from which issued a herd of wild boars. The country through which we were travelling abounded in game. In the higher regions were the ibex, the moufflon, and other kinds of wild sheep and goats. Broad-antlered deer were to be found in the wooded valleys. The red-legged partridge, much larger than that of Europe, everywhere abounded. The 'duroj,' or franeolin, a most delicate bird for the table, swarmed in the brushwood on the banks of the stream.

We had to climb another pass, scarcely less difficult than the one we had crossed on the previous day, before we reached the valley of Borse, through which runs one of the principal confluents of the Karun, here known as the Âbi-Borse (river of Borse). We spent much time in fording it with the animals and baggage—

a matter of no small difficulty, as the water was deep and the stream so rapid that the donkeys could scarcely breast it. We all got more or less wet, and my saddle-bags containing my books and maps were thoroughly soaked. On the opposite side we found a few men who had been left to gather in a scanty crop of barley. They were able to furnish us with some straw for our horses, but nothing whatever, not even dry bread, for ourselves. We were consequently obliged to go to sleep supperless, having eaten nothing the whole day.

When I rose from my carpet at dawn, some of my companions were searching in vain for their shoes, others for their caps. Shefi'a Khan's kaleôn had disappeared. There was scarcely one of us who had not lost something. The men who had furnished us with the provender for our animals were not to be found. They had decamped during the night. After a good deal of vociferation, and much cursing of the thieves, and abuse of their fathers and mothers, we had to resign ourselves to our losses, as there was nothing else to be done. Shefi'a Khan and his people, however, swore that they would never rest until they had inflicted condign punishment upon them. They belonged to the tribe of Dinaruni, notorious even among the Bakhtiyari for being the most audacious and accomplished of thieves. After passing over a thickly-wooded mountain we came upon the encampment of their chief, who, with his followers, was living in huts made of boughs, the usual habitation of the Bakhtiyari during the warm weather. His men were busily employed in building a small fort on the opposite side of the river, which they were constantly crossing and recrossing on inflated sheepskins. They had already completed two circular towers and the curtain or wall uniting them, constructed of rounded stones taken from the bed of the river. The place was called Kessevek. The women, who were unveiled, were employed, seated on the grass in front of their huts, in weaving those carpets so beautiful in colour and design, and so fine in texture, for which the mountains of Luristan are renowned.

Shefi'a Khan lodged a complaint against the thieves who had robbed us during the night. The Khan promised to recover the things stolen, and gave us an excellent breakfast, of which our previous day's fast made us much in need. We then continued our journey, following the Karun by a narrow and very dangerous pathway, led along the precipitous sides of the mountains, where a slip on the part of man or animal would have proved fatal, as the deep stream ran tumultuously below. We stopped for the night, after a short ride, at an encampment of Dinaruni called Bo-hous.

On waking in the morning I found that my quilt had been stolen. This was a severe loss, for although the weather was still mild during the day, the nights were cold, as it was October 3. I was not the only sufferer from the thievish propensities of our Dinaruni hosts. We had another most fatiguing day's journey, scrambling over stony and almost inaccessible mountain ridges, or forcing our way through the thickets of myrtle, oleander, and tamarisk which clothe the banks of the Karun in this part of its course. The mountain slopes were clothed with a kind of heath or heather in full bloom, bearing flowers of the brightest rose colour. Two tracks led to Kala Tul—the castle of Tul—where Mehemet Taki Khan was then residing, as he had left the 'sardisirs,' or mountain pastures, where the Bakhtiyari spend the summer months. One track followed the course of the river and crossed the plain of Mal-Emir, the other took a direct line across the mountains. As the latter would only occupy two days, whilst the former required four, Shefi'a Khan decided upon taking it, although he warned us that it was very bad—even worse than any we had yet followed. This was scarcely credible, but proved true.

We passed through a hamlet called Sheikhun,

surrounded by pomegranate trees in full fruit, but deserted at this time of the year by its inhabitants, who were living higher up on the mountain side. The chief, who received Shefi'a Khan and his followers with the warmest expressions of friendship, embracing them all round, was an immediate retainer of the great Bakhtiyari chief. As he could not persuade them to pass the night in his encampment he insisted that they should remain to breakfast. He slew a sheep for them, and brought us a great bowl of sour milk and delicious honeycombs. As we should have to sleep at the tents of some of his followers who were living in still higher regions, and who would be without the means of properly entertaining us, he insisted upon accompanying us and upon bringing the carcass of a sheep and a great bag of rice with him.

We reached our night's quarters after a most toilsome and dangerous climb, quite justifying Shefi'a Khan's warning. We had now entered the district of Munghast, and had reached a high elevation. The air was keen and piercing, and I had good reason to lament during a bitterly cold night the loss of my wadded quilt.

There was a long and arduous ascent over a treeless and barren slope, and amongst huge rocks covered by mosses and lichen, between which were patches of snow, to reach the summit of the mountain. From this point a magnificent prospect opened before us. Shefi'a Khan pointed out to me the different spots which he considered worthy of note. Almost at our feet could be just seen, like a mere speck, Kala Tul. To the north could be distinguished the plain of Mal-Emir, described to me as containing ruins which excited my liveliest curiosity. Beyond it could be traced the wooded banks of the Karun, as it wound through the low country.

The district of Kala Tul was bounded to the west by a ridge of low, yellow, barren hills, beyond which the imagination rather than the sight might distinguish those vast alluvial plains which stretch, unbroken by a single eminence, to the Euphrates and the sea.

After scrambling and crawling down a most precipitous descent—men and horses appearing to those below them as if piled up one upon the other—we came to a narrow ravine formed by a torrent now dry. Making our way over the loose stones and boulders in its bed we issued into a small plain, and saw, high up on a mound at a short distance from us, the castle of Tul—the end of our long and weary journey.

As we approached, a crowd of men and women came out to welcome their relations and friends who had been so long absent. No news had been received from them during what the Bakhtiyari considered their perilous sojourn within reach of the treacherous and rapacious officers of the Shah and his Government, who were always ready to ill treat and torture his Majesty's subjects, and especially the semi-independent mountaineers. Here and there in the crowd was a young chief mounted on his high-bred horse, with his falcon on his wrist, and followed by his greyhounds. But there was no gathering of armed men. We learned that they were all away with Mehemet Taki Khan, who was absent on the affairs of the tribe at a place called Kala Kùmi. Leaving the rest of my fellow-travellers at the foot of the mound I rode up it with Shefi'a Khan to the castle. As we approached the gateway two younger brothers of the chief and several elders of the tribe, who were seated on raised places on either side of it, rose up to receive us. They exchanged embraces with my companion, and having learnt from him who I was, they bade me welcome and invited me to enter.

I was taken to a large room over the entrance which served as the 'lamerdoun'—as the Lurs call the chamber for the reception of guests. It was already occupied by several persons; but there was a vacant corner in which I could spread

VIEW OF KALA TUL, THE RESIDENCE OF THE BAKHTIYARI CHIEF.

my small carpet upon the felt rug covering the floor. There I deposited the little property that still remained to me. My only spare shirt and stockings, which I had kept in my saddle-bags, had been stolen, and I was thus not even provided with a change of linen. I had every reason to suspect that Khunkiar, the youth whom Shefi'a Khan had attached to me during our journey, was the thief. I complained of him, but without avail. Fortunately my books, maps, and, most important of all, my small assortment of medicines, were still left. My watch and compass, which enabled me to map my route roughly, I had carefully concealed about my person.

Amongst those who occupied the room with me was a seyyid from Shuster—tall, well-featured, and with a long black beard which reached almost to his waist. He had the reputation of being a skilful doctor, and had been invited to Kala Tul to attend one of the children of Mehemet Taki Khan, who was ill of fever. A restless, bright-eyed little man, a native of Isfahan, who was also a physician, had been brought to the castle for the same purpose. He wore the tall lambskin cap, round the lower part of which was twisted a white Cashmere shawl, long robes of silk and fine cloth, a Kerman shawl folded round his waist, with the usual dagger thrust in it, and green shoes with immoderately high heels. There was also in the lamerdoun another seyyid named Kerim, likewise from Shuster—a quiet, mild, studious, and retiring man, who was on a visit to the chief, and who was usually occupied in reading the Koran. A ket-khudâ of one of the tribes, who was at the castle upon business, completed with myself the company that occupied the 'divan-khana,' as it was called. They were my companions for some time. When breakfast and dinner came they ate together from the same dishes, but a separate tray was always brought to me, as I was a Christian, and it was unlawful for them to eat of anything that I had touched.

The three brothers of Mehemet Taki Khan, Au (a Bakhtiyari corruption of Aga) Khan Baba, Au Kerim, and Au Kelb Ali,[2] did the honours of the castle in his absence. The first was a man of prepossessing appearance, with good features, an amiable and rather jovial expression, and gentle address. He was somewhat low in stature and rather stout. Au Kerim was short and thick-set, with a more warlike and determined look than either of his brothers. Au Kelb Ali was tall and thin. His emaciated form and a constant cough showed that he was in an advanced stage

---

[2] They were also called, when addressed in terms of respect, Khan Baba Khan, Kerim Khan, and Kelb Ali Khan.

of consumption. He had, however, a handsome and intelligent countenance, and his manner was more dignified than that of his brothers. They all wore the Bakhtiyari dress with the felt skull-cap.[3]

The chief's wives and family, having only recently descended from the summer pastures in the mountains, were still living in huts of boughs and black tents at the foot of the mound on which stood the castle, where however they removed shortly after my arrival.

Kala Tul resembled the castles that I had already seen in the Bakhtiyari country, being only larger than those belonging to the petty chiefs of the tribe. It was square, with five towers. One of the angles was formed by a square building, in the upper part of which was the 'lamerdoun,' or guest-room. Beneath was the long vaulted passage which formed the entrance to the castle. Within the castle were two courts. In the outer were the rooms for guests and for the chief's immediate attendants and guards; in the inner the women's apartments, or enderun, in which lived the chief and his wives with their maids, and Mehemet Taki's brother, Au Kelb Ali, with his wife. Although this fort, constructed of stone and brick, could have resisted an attack from an irregular force, it

[3] Mehemet Taki Khan had a half-brother named Aslan (Au Aslan, as he was usually called), who also lived at Kala Tul.

could not have been held against troops provided with artillery. On the towers and walls were a few heavy matchlocks eight or nine feet in length on movable stands and turning on a swivel. They were loaded with one large bullet, or with a number of small balls or bits of iron, and were formidable enough to the mountaineers. At the foot of the mound on which the castle stood was a village of mud huts, inhabited by the brothers and relations and retainers of the chiefs in winter. Scattered round it and over the plain were numerous black tents, which were occupied by his dependents who cultivated the soil.

The day after our arrival at Kala Tul, the party which had formed the caravan from Isfahan broke up. Shefi'a Khan with his followers departed for their tents, which were at a day's journey from the castle. The traders with their laden donkeys proceeded on their way to Shuster and other places, to sell the wares they had brought with them.

My reputation as a Frank physician had preceded me, and I had scarcely arrived at the castle when I was surrounded by men and women asking for medicines. They were principally suffering from intermittent fevers, which prevail in all parts of the mountains during the autumn. Shortly afterwards the chief's principal wife sent to ask me to see her son, who, I was told, was dangerously ill, and I was taken to a large booth built of boughs of trees, in which she was living. It was spread with the finest carpets, and was spacious enough to contain a quantity of household effects heaped up in different parts of it. The lady sat unveiled in a corner, watching over her child, a boy of ten years of age, and about her stood several young women, her attendants. She was a tall, graceful woman, still young and singularly handsome, dressed in the Persian fashion, with a quantity of hair falling in tresses down her back from under the purple silk kerchief bound round her forehead. As I entered she rose to meet me, and I was at once captivated by her sweet and kindly expression. She welcomed me in the name of her husband to Kala Tul, and then described to me how her son had been ill for some time from fever, and how the two noted physicians whom I had seen in the lamerdoun had been sent for from a great distance to prescribe for him, but had failed to effect a cure. She entreated me, with tears, to save the boy, as he was her eldest son, and greatly beloved by his father. I found the child in a very weak state from a severe attack of intermittent fever. I had suffered so much myself during my wanderings from this malady that I had acquired some experience in its treatment,

I promised the mother some medicine and told her how it was to be administered. Returning to the castle I sent her some doses of quinine; but before giving them to the child she thought it expedient to consult the two physicians who had been summoned to Kala Tul. Fearing that if their patient passed into my hands they would lose the presents they expected, they advised that it would be dangerous to try my remedies. Their opinion was confirmed by a mulla, who, upon all such important occasions, was employed to consult the Koran in the usual way by opening the leaves at random. The oracle was unfavourable, and my medicine was put aside for the baths of melon juice and Shiraz wine, and the water with which the inside of a porcelain coffee-cup, on which a text from the Koran was written in ink, had been washed. The condition of the boy, however, became so alarming that his father was sent for.

Soon afterwards Mehemet Taki Khan arrived at the castle, surrounded by a crowd of horsemen. Leaving them at the foot of the mound, he rode up to the entrance, and dismounting from his mare—a magnificent Arab of pure breed—seated himself on the raised platform of masonry where the chiefs and the 'rish-sufids'[4] usually assembled in the

---

[4] *I.e.* 'white-beards,' a title given to the elders of a tribe or village.

afternoon and in the evening to talk over the events of the day, to listen to complaints, and to settle disputes. It was, as it were, the judgment-seat of the tribe, whence justice was administered, redress given, and punishments awarded. The elders acted as assessors to the chief, who was all-powerful, and exercised the right of life and death over his people.

The guests at the castle, myself included, came down to greet him. I presented my firman and the letter with which the Matamet had furnished me. He glanced at them, and then threw them somewhat contemptuously from him. This reception was not very encouraging, and I began to fear that my presence was not welcome to the Bakhtiyari chief, and that he might entertain suspicions as to the object of my visit to Kala Tul. But my misgivings were soon removed. Motioning me to be seated, and addressing me in a very friendly tone, he said that I was in no need of such an introduction as I had brought to him, or of the firman of the Shah, which had no authority among the independent mountain tribes. As a stranger I was welcome to his house, which I was to consider as my own as long as I liked to remain at Kala Tul. He added that he had already received accounts of me from his vizir, Shefi'a Khan, and that he considered my arrival as of happy augury for himself and his people.

The cordial and unaffected manner in which these words were spoken made a very favourable impression upon me, which was confirmed by his frank and manly bearing and engaging expression. Mehemet Taki Khan was a man of about fifty years of age, of middle height, somewhat corpulent, and of a very commanding presence. His otherwise handsome countenance was disfigured by a wound received in war from an iron mace, which had broken the bridge of his nose. He had a sympathetic, pleasing voice, a most winning smile, and a merry laugh. He was in the dress which the Bakhtiyari chiefs usually wore on a journey, or when on a raid or warlike expedition—a tight-fitting cloth tunic reaching to about the knees, over a long silk robe, the skirts of which were thrust into capacious trousers, fastened round the ankles by broad embroidered bands. Round his Lur skull-cap of felt was twisted the 'lung,' or striped shawl. His arms consisted of a gun, with a barrel of the rarest damascene work, and a stock beautifully inlaid with ivory and gold; a curved sword, or scimitar, of the finest Khorassan steel — its handle and sheath of silver and gold; a jewelled dagger of great price, and a long, highly ornamented pistol thrust in the 'kesh-kemer,' or belt, round his waist, to which were hung his powder-flasks, leather pouches for holding bullets, and various objects used for priming and loading his gun, all of the choicest description. The head and neck of his beautiful Arab mare were adorned with tassels of red silk and silver knobs. His saddle was also richly decorated, and under the girths was passed on one side a second sword, and on the other an iron inlaid mace, such as Persian horsemen use in battle. Mehemet Taki Khan was justly proud of his arms, which were renowned throughout Khuzistan. He had a very noble air, and was the very beau-ideal of a great feudal chief.

Mehemet Taki Khan, by his courage and abilities, had raised himself from the rank of a chief of the clan of Kunursi, to which he belonged, to that of head of the great Bakhtiyari tribe of Chehar-Lang, of which the Kunursi were a branch. He was descended from Ali Mardan Khan, a Bakhtiyari chief, who, during the anarchy that prevailed after the death of Nadir Shah, possessed himself of Isfahan and caused himself to be proclaimed king of Persia—an honour which he enjoyed but for a short time, having been put aside by Kerim Khan Zend. Of this lineage he was very proud, but it made him an object of jealousy and suspicion to the Persian Government, and encouraged him in his desire to establish his independence.

He was famed throughout Persia, as well as in his mountains, as a dauntless warrior, a most expert swordsman, an excellent shot, and an unrivalled horseman. He was not less celebrated for his skill in dealing with tribal politics, and as an administrator of tribal affairs. His rivals, and those who attempted to dispute his authority, had one after another been overcome and reduced by him to submission, or had been slain in war or by treachery. Ali Khan, his father, was too powerful not to excite the suspicion of the Persian Government. He was betrayed by his brother, Hassan Khan, and Feth Ali Khan, his uncle, into the hands of the Shah. His eyes, according to the barbarous custom of the country, were put out, and Hassan Khan received the chieftainship as the reward of his perfidy. Mehemet Taki Khan and his brothers were then children, and were concealed in the Armenian village of Feridun. Hassan Khan, to strengthen his authority over the tribes, put to death Iskander Khan, an uncle of Mehemet Taki Khan, with two other of his nearest relations, and attempted to slay his brother and his two sons. According to the law which prevails among the nomad tribes, the blood of Hassan Khan and two of his family was required by Mehemet Taki Khan. He and his brothers, Ali Naghi and Khan Baba, in order to avenge the murder of their relatives, succeeded in penetrating without discovery into the castle of Hassan Khan, and slew him as he rose from his prayers. Mehemet Taki Khan subsequently married the daughter of Hassan Khan, and adopted his three infant sons, with a view to putting an end to the blood feud and consequent dissensions which had led to a war between the two branches of the tribe. In this he was successful, and he had for some time maintained peace in the mountains over which he ruled, although he could not altogether prevent petty chiefs from lifting each other's cattle, which led to quarrels frequently ending in bloodshed.

Although tribal politics in Asia are notoriously tainted with, if not founded upon, treachery and deceit, Mehemet Taki Khan had the reputation of being a generous and merciful enemy, and a trustworthy, just, and humane man, and his followers were devotedly attached to him. He could neither read nor write, but he was exceedingly intelligent, and especially fond of poetry. He was sincerely anxious to promote the good of his people and the prosperity of his country by maintaining peace, by securing the safety of the roads through his territories, and by opening his mountains to trade.[5]

[5] Sir Henry Rawlinson, who took part as an officer in the Persian army in an expedition against Mehemet Taki Khan, wrote

He had scarcely entered the enderun of the castle, to which his wife had removed, than he sent for me. I found him sobbing and in deep distress. His wife and her women were making that mournful wail which denotes that some great misfortune has happened or is impending.[6] The child was believed to be at the point of death. The father appealed to me in heartrending terms, offering me gifts of horses and anything that I might desire if I would only save the life of his son. The skilful physicians, he said, for whom he had sent, had now declared that they could do nothing more for the boy, and his only hope was in me.

I could not resist Mehemet Taki Khan's entreaties, and after reminding him that the medicines I had already prescribed had not been given,

of him in the following terms: 'At the outset of his career he was the acknowledged chief of his own single tribe, and he owes his present powerful position solely to the distinguished ability with which he has steered his course amid the broils and conflicts of the other tribes. The clans, one by one, have sought his protection, and enrolled themselves amongst his subjects; and he can now at any time bring into the field a well-armed force of 10,000 or 12,000 men. He collects his revenues according to no arbitrary method, but in proportion to the fertility of the districts and prosperous state of the villages and tribes. He has done everything in his power to break the tribes of their nomadic habits, and to a great extent he has succeeded.'—*Notes on a March from Zohab to Khuzistan*, p. 105.

[6] It consists of the constant repetition of a plaintive sound like 'Wai, wai,' whilst the body is rocked to and fro.

I consented to do all in my power to save the child's life, on condition that the native doctors were not allowed to interfere. Although he was willing to agree to all I required, he could not, as a good Musulman, allow the boy to take my remedies until the mulla, who resided in the castle and acted as secretary and chaplain to the chief, could consult the Koran and his beads. The omen was favourable, and I was authorised to administer my medicines, but they were to be mixed with water which had served to wash off from the cup a text from the Koran—a ceremony upon which the mulla insisted.

The child was in a high fever, which I hoped might yield to Dover's powder and quinine. I administered a dose of the former at once, and prepared to pass the night in watching its effect. I was naturally in great anxiety as to the result. If the boy recovered I had every reason to hope that I should secure the gratitude of his father, and be able to carry out my plan of visiting the ruins and monuments which were said to exist in the Bakhtiyari Mountains, and which it was the main object of my journey to reach. If, on the other hand, he were to die, his death would be laid at my door, and the consequences might prove very serious, as I should be accused by my rivals, the native physicians, of having poisoned the child. About midnight, to my great relief, he broke out into a violent perspiration, which all the remedies hitherto given him had failed to produce. On the following day he was better. I began to administer the quinine, and in a short time he was pronounced out of danger, and on the way to complete recovery.

The gratitude of the father and mother knew no bounds, for the affection among these mountaineers for their children is very great. They insisted that I should in future live in the enderun, and a room was assigned to me. Mehemet Taki Khan made me accept a horse, as mine had not recovered from the effects of the journey over the mountains. But what I most needed was linen and clothes. These were supplied to me by his wife. I was indeed sadly in want of my second shirt. I had been compelled, after I had been robbed of it, to hide myself in the rushes on the bank of a stream to wash the one I wore, and to wait without it until it had been dried by the sun. My Persian clothes, of European cotton print, were in the shabbiest condition, and beyond repair. The Khatun's women soon made for me all that I was in want of.

Khatun-jan—'Lady of my soul'—was the principal wife of Mehemet Taki Khan, and the mother of his three children.[7] There were two other ladies who ranked as wives of the chief, but who were on a very different footing from the Khanum, whose apartment her husband regularly shared. She was one of the best and kindest women I ever knew. She treated me with the affection of a mother, nursing me when I was suffering from attacks of fever, which were frequent and severe, and during which I was frequently delirious for several hours. She took charge of the little money that I possessed, as she feared that in my wanderings in search of ruins and inscriptions I might be exposed to great danger if it were known that I carried it with me. She acted as my banker, and gave me what I needed for immediate use, which was very little indeed, as there was nothing to buy, all that I required being furnished to me by her husband and herself. Neither she nor her women, nor indeed any of the wives and female relatives of the chief and his brothers, ever veiled themselves before me. I was in the habit of passing the evening listening to the Khanum's stories about the tribes. The chief was frequently present and took part in the conversation. I was even permitted, contrary to the etiquette of the

---

[7] Her full name was Khatun-jan Khanum. Khatun and Khanum have the same meaning, that of Lady, and the name translated would be 'Lady lady of my soul.'

harem, to eat with her, and Mehemet Taki Khan would jokingly taunt me with introducing European customs into the enderun, as it was not proper for even the husband to sit at the same tray with his wife, although in private. The other wives of the Khan, who were young and not ill-looking, never sat in his presence unless invited to do so, taking their places among the waiting-women of the Khanum, who was always treated with the greatest respect and consideration by her husband, and by her partners in his affections. She was the daughter of one of the principal chiefs in Luristan, and consequently a lady of rank and entitled to this treatment. Mehemet Taki Khan, in speaking of her, always called her 'the mother of Hussein Kuli'—her eldest son, my patient—and not by her name.

Khanumi, Khatun-jan's sister, who was some years younger than herself, was the beauty of Kala Tul. Indeed, it was said that there was not a more lovely woman in the tribe, and she deserved her reputation. Her features were of exquisite delicacy, her eyes large, black, and almond-shaped, her hair of the darkest hue. She was intelligent and lively, and a great favourite with all the inmates of the enderun. The chief and the Khanum would often tell me that if I would become a Musulman and live with them they would give her to me for wife. The inducement was great, but the temptation was resisted.

Khatun-jan's mother, Fatima, was another of my friends in the harem, and showed me at all times great kindness. She was still young, and had not lost her figure or good looks, as women among nomad tribes usually do after they have borne children. She had a rich store of tribal histories and traditions, and would relate in vivid and picturesque language the wars of the Bakhtiyari, their blood feuds, and those barbarous deeds of revenge which stain their annals.

Hussein Kuli was the eldest of Mehemet Taki Khan's three sons by Khatun-jan. Both father and mother doted on him. He was one of those beautiful boys who are constantly seen in Persia, and especially among the mountain tribes, and was intelligent, high-spirited, brave, and dauntless—inheriting all the qualities of his father. He became greatly attached to me and I to him. His two younger brothers were also charming children. The elder of the two, Meta-Kuli, was familiarly called 'Berfi,'[8] from having been born in the 'yilaks,' or summer quarters, of Zerda Kuh, among the snowfields. The other was named Riza-Kuli.

Mehemet Taki Khan's youngest brother, Au Kelb Ali, the tall, emaciated youth who had

[8] *I.e.* of the snow.

received me on my arrival at the castle, also lived in the enderun. He was lovingly tended by his one wife, who watched him night and day with the greatest devotion and care as he gradually sank under the disease from which he was suffering. I could do nothing for his relief, notwithstanding her entreaties.

The chief's other brothers, Ali Naghi Khan, Au Khan Baba, and Au Kerim, lived in the village with their families. I was intimate with them all, and had free access to their enderuns. Some of their wives were very handsome women. The eldest of the three brothers was Ali Naghi Khan, whom I had seen at Isfahan, and who had been sent to Tehran as a hostage for the loyalty of Mehemet Taki Khan. He was known among the Bakhtiyari as the 'Serdar,' or commander-in-chief, and being considered the ablest of the chiefs, was usually sent upon political and diplomatic missions to the capital and to the governor of Isfahan. He had consequently seen something of the Court and of Persian officials, who were notorious for their profligacy. He had unfortunately acquired most of their vices, and especially a fondness for those drunken orgies by running water and amidst flowers, with the accompaniment of music and of the amorous couplets of Saadi and Hafiz in which they were wont to indulge. To these pleasures neither Mehemet Taki Khan, who was a rigid Musulman and very strict in his religious observances, nor his other brothers were given. He never drank wine or arak, and they were forbidden in the castle, although occasionally a debauched Persian would indulge in them in the guest-room, to the great horror and scandal of the pious mullas and seyyids who frequented it.

Ali Naghi Khan had three wives, two of whom had accompanied us from Isfahan. They were living in spacious booths, built of the boughs of trees and reeds, and divided into several compartments spread with the finest carpets, and furnished with such luxuries as the chief of a nomad tribe could procure. They were all three very beautiful women. One, whose name was Bibi Limûn (the Lady Lemon), was the daughter of a Bakhtiyari chief; the others were Georgians whom he had purchased. They lived together, apparently in perfect harmony, and had the same attendants. Their husband, like other Bakhtiyari chiefs, had but one establishment, and occupied with them one hut, or when encamping in the mountains one large tent, divided into four compartments; one was reserved for guests; in another were picketed at night the favourite horses of the chief; in a third were kept the caldrons, cooking and baking utensils, great bundles of bedding, and other furniture. The last to the right was occupied by the ladies and their women.

The domestic arrangements were very simple. Each person had his or her wadded quilt or coverlet, a small carpet, and a bolster, which during the day were rolled up in a silk or linen cover. When the time for going to rest had come, the tent was closed by lowering the flaps in front, which were held up during the day by poles, except in the summer, when in the suffocating heat of the arid valleys of Khuzistan, it was left open on all sides. The bundles containing the bed-clothes were then unrolled, the carpets spread either on the bare ground or on a 'nemud,' or rug of soft felt—generally in hot weather outside the tent—and every one settled himself or herself for the night. This did not occupy much time, as neither women nor men did more than take off the outer coat or jacket, loosen the strings or buttons which fasten the shirt and long robe in front, and remove the shawl and belt from round the waist. In the morning the men, crouching down at a short distance from the tent, performed their ablutions, which consisted of washing the face, hands, and arms with water, poured into the open palm of the right hand from a jug of elegant form with a long curved spout, rinsing their mouths, and rubbing their teeth with the forefinger, also of the right hand, the left never being used for such purposes or in eating. They then put on the few garments they had taken off before going to rest, said their morning prayer, and were ready for the day.

The women performed their toilettes inside the tent, which, however, was generally open on one side. But their preparations differed little from those of the men, as they had only to put on their tunics and jackets. Their dress was nearly similar to that of the Persian women, but, except in the case of the wives of the chiefs, made of coarser materials woven by themselves, or of common European chintzes bought of itinerant pedlars. It consisted of the ample loose shalwars, or trousers, of chintz or red silk, tied above the hips and very full at the ankles; a short chemise of white linen only reaching to the band of the shalwars, entirely open in front, but fastened with a loop at the neck; and a jacket, usually of common European figured chintz, but occasionally of silk, also open in front, fitting tight to the arms as far as the elbow, and then hanging loosely. Sometimes in winter an outer tunic of cloth of similar shape was worn. The jackets of the wives of the chiefs were of Cashmere shawl, or of silk or velvet, frequently embroidered with gold. The whole of the bosom and the rest of the person to the waist were exposed; but when receiving strangers, and sometimes before their husbands as a mark of respect, the ladies tied a large silk kerchief, usually of a rich plum colour, round their necks, which concealed the throat, chest, and arms. Their hair fell in tiny plaits down the back, and was gathered in curls and ringlets on the forehead and on the side of the face. A kerchief of black silk, or of white linen in the case of the poorer people, was bound round the head, the ends being left to hang loosely behind. In the enderun the ladies sometimes wore skull-caps of Cashmere shawl, ornamented with pearls and jewels. The Bakhtiyari women rarely wore stockings, and generally used woollen shoes, knitted by themselves, with leather soles; but sometimes they put on the Persian shoes of green leather with very high heels, which were fashionable in the towns. They were fond of ornaments, wearing bracelets, armlets, anklets, and necklaces of gold or silver, and they invariably carried, either hung round the neck, bound round their arms, or attached to some part of their dress, amulets or charms consisting of texts from the Koran written on parchment and enclosed in a small silver case.[9]

[9] The men wore similar charms. Mehemet Taki Khan and other chiefs carried the entire Koran, written in very minute characters, suspended round their necks in an embossed silver or embroidered leather case. Such manuscripts, when of very small size, are greatly prized.

When the Lur women went abroad among strangers they enveloped themselves in an ample veil, which concealed the whole person. and was held over the face. But in their tents, and when moving about in their encampments, they did not conform to the Musulman custom of hiding their features.

Like the Persians, the Bakhtiyari men and women, when they could afford the luxury, dyed their hair, their eyebrows, the palms of their hands, the soles of their feet, and the nails of their fingers and toes. This was done in the bath, when there was one to be found, which was rarely the case in the mountains except in some of the castles of the chiefs. The following was the process, to which I had to submit once a week, besides having the centre part of my head shaved without soap and with a rasping razor which brought tears into my eyes. The dried leaves of the henna were made into a paste with water. A little lime-juice or some other acid was added. This paste was laid over all the parts to be dyed for about an hour, when it was washed off. The colour produced was of a dark red approaching to brown. It was thus left on the hands, feet, and nails; but the hair and eyebrows were then covered for another hour with a second paste made of the leaves of the indigo plant, which turned the red-brown into a glossy black.

The wives of the chiefs, like the women in the towns, rubbed their eyelids with the black powder called 'kohl,' which added to the brilliancy of their eyes and to their fascination, and, like our great-grandmothers, adorned their cheeks with black patches or beauty-spots. Both men and women used depilatories.

The costume of the Bakhtiyari women was not unbecoming. The exposure of so much of the person was not in accordance with European notions of propriety; but it was not considered otherwise than strictly decent by women who would carefully conceal their throats and hide their faces from a stranger, and would be greatly shocked by the low dresses worn by European ladies. What women may leave uncovered and what they must keep carefully concealed not to offend modesty, is a question of habit and fashion. I have seen Arab girls on the banks of the rivers of Mesopotamia raise their one solitary blue shirt and bring it over their heads to hide their faces from a European.

The Bakhtiyari, like nearly all independent and proud mountain tribes, are exceedingly sensitive about their women. An insult or outrage to a man's wife or female relative could only be avenged by the death of the person guilty of it.

MARRIAGES

Even in their tribal wars and quarrels, where the lives of men are freely taken, the honour of a woman is respected. A wife who had betrayed her husband would, if discovered, be almost invariably slain, as would her paramour. The security enjoyed by women enables them to travel alone, without fear of harm, in those wild mountains where every man holds his life in his hand.

Among the Bakhtiyari, as among other Eastern peoples, marriages are contracted at a very early age. Children are affianced from their tenderest years, and boys are frequently married when fourteen or fifteen years old, the girls when twelve or even before. It is rare to find any but the chiefs with more than one wife. Divorce is so easy among most Mohammedans that it can be effected by the repetition of a few words by the husband. That curious law, or custom, called 'sigha,' prevailing among the Shi'a Musulmans of Persia, which enables a man to marry for a period however short—even for twenty-four hours—and which makes the contract for the time legal, and according, they assert, to the precepts of the Koran, prevails in the mountains of Luristan, although it is rarely had recourse to. It is founded upon the facility of divorce. A man contracts a marriage with a woman before a mulla. according to the prescribed forms of the Mohammedan law, on an understanding on her part that she is to be divorced after a certain period, whether long or short; strangers thus enter into legal marriages with women whom they do not wish to take with them to their own country, or who do not themselves desire to leave their families and native place. There were mullas in Isfahan, and in other large cities and towns in Persia, who made a livelihood by celebrating marriages of this nature. The profession is not, however, considered a very respectable one.

But to return from this digression. Other members of Mehemet Taki Khan's family lived in the village. I knew all of them, and was received by them and their wives as a welcome guest. I was frequently able to render them little services by prescribing for them or their children, who were suffering from fever or sore eyes. Happily, my treatment generally proved successful, and my reputation as a physician consequently increased.

The following was our mode of life in the castle. In the morning trays with dishes of excellent pillaus of rice and boiled mutton, cooked by the women of the harem, and bowls of sherbet, of sugar and water, flavoured with some kind of syrup, were brought into the diwân-kâna for the guests. The chief himself usually breakfasted in the enderun. I did as I pleased.

After breakfast Mehemet Taki Khan left the women's apartments and seated himself on the raised brick platform at the entrance to the castle, where he was joined by some of the notables and 'rish-sefids.' He there heard and settled disputes, administered speedy justice, or received travellers bringing news from afar, or messengers with letters upon business and public affairs. Later in the day he would order his favourite horses, of which he had ten or twelve always tethered in the inner court of the castle, where they were tended with the greatest care, to be brought out to be inspected. They were of the finest Arab breeds—Wusnan, Suglowiyah, Kailan, and others—and he was very proud of them. They had been either obtained from the Arab tribes on the Euphrates, or had been bred by him from horses of the best races of Arabia. He knew the genealogy of all of them. He usually mounted one of them whilst the rest were being exercised by his attendants, who galloped to and fro in the plain or wheeled in narrowing circles, discharging their guns, like the Parthians of yore their arrows, from behind as they fled from an imaginary foe, picking up a handkerchief or other object when at full speed, and performing other feats, such as hitting with a single ball their felt skull cap which they had thrown on the ground, and clinging at full length to one side of their horse in order not to offer a mark to the enemy. Mehemet Taki Khan's horsemen were considered the most skilful and daring in Persia.

A favourite amusement of the chief was to exercise his horses to the chase, by bringing them up to a rudely stuffed lion which was kept for the purpose in the castle. They were thus accustomed to the sight and smell of this animal, which is frequently found in the valleys and plains of Khuzistan, and which is often hunted by the Bakhtiyari.

I frequently accompanied the Khan's brother, Au Kerim, who was an ardent sportsman, and other young chiefs, with their hawks and their greyhounds, on hunting expeditions. The plain of Tul and the neighbouring valleys abounded with the large red-legged partridge, and the duroj, or black partridge. The 'hubara,' or middle-sized bustard, was also constantly met with, and in the marshy ground near the streams ducks and other water-fowl were plentiful. Hawks, trained to hunt with the large, long-haired Persian, and the more high-bred Arab, greyhound, were used for the capture of hares and gazelles.

We occasionally ascended the mountains behind Kala Tul, and rarely returned without two or three ibex, of which we usually saw large herds, or a moufflon or mountain sheep The flesh was generally cut up into small bits, spitted on a ramrod or skewer, and roasted as 'kibabs.' The liver, heart, and other parts of the entrails thus cooked were greatly relished by the mountaineers.

At sunset attendants bearing trays on their heads appeared in the lamerdoun. The dinner consisted of the usual pillaus, with the addition of kibabs, stewed fowls, roast game, and several kinds of sweet dishes. After dinner coffee was handed round in the Arab fashion, kaleôns were smoked, and some of the guests played at backgammon, whilst others conversed or read or recited poetry until it was time to sleep, when every one spread his carpet upon the floor and settled himself for the night. I usually dined in the enderun. Mehemet Taki Khan was fond of talking with me about England and her institutions and European inventions. He took a very enlightened view of such matters, was eager to induce the wild inhabitants of his mountains to engage in peaceful pursuits, and was very desirous that the country should be opened to commerce. These conversations generally took place in the evening in the inner court, where his favourite horses were tethered, and where he would sit amongst them on his carpet. But he was also in the habit of questioning me on those subjects when we were seated at the entrance to the castle, surrounded by the elders and principal men of the tribe. He would make me describe to them railways and various modern discoveries, and explain to them the European sciences of astronomy, geology, and others unknown to his people. As they were at variance with the teachings of the Koran, he would direct a mulla to argue the matter with me and to endeavour to confound me. The learned man was generally satisfied with a simple denial of what I had stated, quoting in support of it some verse from the holy volume. But this did not satisfy the chief, who was anxious for knowledge. He would make me describe the wigs worn by judges and barristers in England, and then, with a jovial laugh, would exclaim, 'You see that to make a cadi in England it only requires two horses' tails!' He had some difficulty in understanding why I had left my home to incur the privations and dangers of a journey through wild and inhospitable regions. He could scarcely believe that I had been impelled to do so by the love of adventure, and by a curiosity to visit new countries and to explore ancient remains. It was not easy to remove his suspicion that I was a secret agent of the British Government, travelling to obtain topographical and other information with a view to the invasion of Persia by England, as the news of the rupture of diplomatic relations between the two countries had already reached the Bakhtiyari Mountains. But he so hated the corrupt, vicious, and cruel Persians, and was so exasperated at the constant demands upon him for money from the Persian governor of Isfahan, that he was not the less friendly to me on that account.

The lessons in Persian that I had taken at Baghdad, and the necessity of using it when travelling alone, had enabled me to speak the language with some fluency, although, of course, incorrectly. Whilst at Kala Tul, Seyyid Kerim, whom I had found in the guest-room on my arrival there, and who had made a very favourable impression upon me by his quiet and reserved manners and his amiable expression, undertook to give me further instruction in it, and I read with him the sonnets of Saadi and Hafiz and parts of the 'Shah-Nameh.' He also taught me to write the Persian character, and the knowledge that I acquired of it proved very useful to me.

The Bakhtiyari speak a Persian dialect which is generally known as the Luri, and is a corruption of the pure old Persian without the modern intermixture of Arabic and Turkish. They maintain, indeed, that it is the 'Farsi Kadîm,' the language of the ancient Persians. It more nearly resembles the language of the 'Shah-Nameh' than it does that of the works of the later Persian poets and of modern Persian literature. I was soon able to converse in it; but it had the effect of corrupting the little Persian that I had originally learnt, and my Persian friends laughed at me because I used Lur words and expressions.

The Bakhtiyari are probably the descendants of the tribes which inhabited the mountains they still occupy from the remotest antiquity. They are believed to be of pure Iranian or Persian blood. They are a splendid race, far surpassing in moral, as well as in physical, qualities the inhabitants of the towns and plains of Persia—the men tall, finely featured, and well built; the women of singular beauty, of graceful form, and when young almost as fair as Englishwomen. If the men have, for the most part, a savage and somewhat forbidding expression, it arises from the mode of life they have led from time immemorial. They are constantly at war, either among themselves or with the Persian Government, against which they are in chronic rebellion. In addition, they are arrant robbers and freebooters, living upon the plunder of their neighbours or of caravans, or of the pusillanimous population of the plains, amongst which they are in the habit of carrying their forays with impunity. But notwithstanding the fierce and truculent appearance of the men, I have never seen together finer specimens of the human race than in a Bakhtiyari encampment.

The Bakhtiyari tribes have at different times played an important part in the history of Persia. Their chiefs would descend into the plains at the head of large bodies of brave and daring horsemen. Sometimes they threatened Isfahan, the capital; at others they encountered the enemies of their country, such as the Afshars, who had overrun the greater part of it. They rebelled against the renowned Nadir Shah, who, however, conquered them, and removed some of their tribes to distant parts of his empire. But during the anarchy which prevailed after his death they revenged themselves by seizing the throne and proclaiming Ali Mardan Khan—who, as I have mentioned, was the ancestor of Mehemet Taki Khan—Shah of Persia.

CHAPTER IX.

Excursion to Mal-Emir—Bakhtiyari graves—The Atabegs—A wife of Mehemet Taki Khan—Plain of Mal-Emir—Mulla Mohammed—Sculptures and inscriptions of Shikefti-Salman—Leave Mal-Emir for Sûsan—Robbed on the road—An Iliyat encampment—Difficulties in crossing the Karun—Mulla Feraj—The tomb of Daniel—A fanatic—Suspicions of the Bakhtiyari—The ruins—Ancient bridge—Bakhtiyari music—Leave Sûsan—Forest encampment—Return to Kala Tul—Recover my property—Visit ruins of Manjanik—Legend relating to Abraham—Ill of fever—Village of Abu'l Abbas—Attempt to visit Shefi'a Khan—Dangers of the road—Return to Kala Tul—Accompany Shefi'a Khan to his tents—A terrible night—Encounter with a lion—The lions of Khuzistan—Leopards and bears—Recalled to Kala Tul—Escape from drowning.

DURING the autumn months I made excursions in search of ruins and inscriptions which were said to exist in the neighbourhood of Kala Tul. My attention had been directed by Sir Henry Rawlinson's memoir on Susiana to several sites in the Bakhtiyari Mountains where ancient remains were believed to exist. He had not visited the country himself, and his account of them had been derived from Lur chiefs whom he had met. Such accounts are not to be relied on when coming from ignorant Orientals, who are at all times disposed to exaggerate very greatly what they may have seen, in order to excite the wonder and curiosity of Europeans, and are, moreover, incapable of accurately describing such things, even when desirous of doing so in good faith. I was led more than once to make difficult and perilous expeditions without any result whatever. The great temples and palaces, and the rock-cut inscriptions of which Sir Henry Rawlinson had received the most minute descriptions, generally proved to be a few heaps of stone, or the ruined walls of a building of a comparatively recent period, and some natural marks on a weather-beaten cliff, in which a lively imagination had detected the writing of the Franks.

One of my first expeditions from Kala Tul was to the plain of Mal-Emir,[1] and to a valley known to the Bakhtiyari by the name of Sûsan, or Shushan. Sir Henry Rawlinson had been informed that extensive ruins and inscriptions cut in the rock existed at both places, and that at Sûsan there was a tomb which was traditionally known as that of the prophet Daniel. He had been consequently led to infer that these ruins represented the ancient city of Shushan, in the province of Elam and on the river Ulai, where, in the palace, Daniel saw his

---

[1] *I.e.* the house or treasure of the prince; 'mal' may signify either.

vision.² Some mounds also known as Sûsan, and a building held sacred by Jews and Mohammedans as the tomb of the prophet, on the small river Shapur, or Shaour, near the modern town of Dizful, had been visited and described by an English traveller, Colonel Monteith,³ and were generally believed by geographers to mark the site of the Susa of the Greeks—the capital of the ancient kingdom of Susiana or Elymais. There being thus two places within the boundaries of ancient Susiana called Sûsan, and there being moreover a traditionary tomb of Daniel in both of them, Sir Henry Rawlinson had endeavoured to explain the fact by conjecturing that there had been two cities of the same name, one of which was distinguished as Shushan the palace, and was to be identified with the ruins said to exist in the Bakhtiyari Mountains. It was accordingly of some importance that the seruins should be examined.

Mehemet Taki Khan being absent from the castle, his brother, Au Khan Baba, offered to give me letters for the chiefs of two small tribes which were encamped in the districts which I wished to visit. He warned me, however, that they were a

---

² Daniel viii. 2. 'And I saw in a vision; and it came to pass, when I saw, that I *was* at Shushan *in* the palace, which *is* in the province of Elam; and I saw in a vision, and I was by the river of Ulai.'

³ *Travels in various Countries of the East*, vol. ii. p. 426.

very lawless set of fellows and most notorious robbers, and that, as they neither respected the laws of hospitality nor the precepts of the Koran, I should run no inconsiderable risk in trusting myself amongst them, even although under the protection of Mehemet Taki Khan. At that time I was not so generally known amongst these wild mountaineers as I afterwards came to be, as the guest and friend of their great chief. But I was resolved not to lose the opportunity of exploring these remains now that I was within reach of them.

Accordingly, one morning I mounted my horse, and, accompanied by a guide and provided with letters for the two petty chiefs, who were named Mulla Mohammed and Mulla Feraj, left the castle for the plain of Mal-Emir. I had deposited the little money I possessed with Khatun-jan Khanum. She persuaded me to leave in her charge my double-barrelled gun, which I still retained, as the sight of it might expose me to danger—firearms being greatly prized. She was anxious, too, that I should let her keep my watch and compass, which were likely to excite the cupidity of the people among whom I was going. But as they were necessary to me for making observations and mapping my route, I would not part with them.

After crossing the plain of Tul I reached a range of low hills, which divides it from that of

Mal-Emir. Before entering them I passed an 'Imaum-Zadeh,' or tomb of a Musulman saint, surrounded by a large number of graves of persons whose relatives had brought their remains to be interred near those of the holy man. Those of chiefs and noted warriors were marked by headstones upon which were rudely sculptured the figure of a lion, and such emblems of prowess in war or in the chase as a gun, a sword, a spear, and a powder-flask. Those of the women had carved upon them some object of female use, such as a comb. Women wailing for their husbands or children were seated near some of the graves, rocking themselves to and fro, uttering their long, melancholy wail, tearing their hair, scratching their faces with their nails, and beating their naked bosoms. I rarely passed a Bakhtiyari burial-place without seeing women thus engaged, as they continued to mourn for long in this way for those whom they had lost.

In the hills I passed several remains of ancient buildings, which my guide said had belonged to the Atabegs, who formerly held sway in Luristan, and to whom all ruins in this part of the country are generally attributed. A spring at which we stopped was called 'Chesmeh Atabeghi,' or the Atabegs' spring. We arrived in the afternoon at a village named Alurgon, surrounded by pomegranate and fig-trees in full fruit, and by rice-fields. Near it were the ruins of a castle. It had formerly belonged to Hassan Khan, a powerful chief of the Char-Lang tribe, who, as I have mentioned, with his brother, Feth Ali Khan, and one of his sons, had been slain by Mehemet Taki Khan, in revenge for their treachery in delivering into the hands of Feth Ali Shah his father, Ali Khan. Hassan Khan's daughter, who had been married to Mehemet Taki Khan in order to bring the blood-feud to an end, was from her rank considered among the tribes as his principal wife, but they did not live together. She resided in the ruined castle, but was too ill with intermittent fever to see me. She sent me sweetmeats and fruit, and a pillau and other dishes for supper. The elders of the village came to me in the evening, and, sitting round the fire, related to me the story of Hassan Khan's death — how Mehemet Taki Khan had fought with him hand to hand, and how he had slain Feth Ali Khan, pointing out the spot where he fell.

A narrow gorge, at the entrance of which were two ruined towers, apparently of the Sassanian period, and once intended for its defence, led into the plain of Mal-Emir. The chief, Mulla Mohammed,[4] for whom I had a letter from Au Khan

[4] Some of the chiefs of the petty tribes in this part of the

Baba, was encamped among a number of artificial mounds marking the site of an ancient city of some size. The largest of them, about forty feet in height, was called the 'kala,' or castle.

Mulla Mohammed and his followers were living in huts constructed of reeds and boughs. I was received in the place reserved for guests, and delivered my letter from Au Khan Baba; but when I asked to be shown the inscriptions cut in the rock, which I had been told were not far distant, they seemed disposed to throw difficulties in my way. They wished to know why I desired to visit them. Was I in search of treasure? Were the Feringhi about to return to take the country? Finding that I was determined to see the inscriptions, which they admitted were not far off, whether they wished it or not, Mulla Chiragh, the brother of Mulla Mohammed, with two men, volunteered to accompany me. They led me to a narrow gorge in which was a large cavern containing a natural recess, on either side of which was a figure, much larger than life, sculptured in the rock. The one to the right, with a long curled beard, appeared, from the head-dress or cap fitting close to the head with a double fold over the fore-

---

Bakhtiyari country have this title of 'Mulla,' which does not imply that they are men of the law or priests.

head, to be that of a 'mobed,' or priest of the ancient fire-worshippers. His robe reached to his feet, and his arms were folded on his breast. The other figure had a similar head-dress, but wore a short tunic, and his hands were joined in an attitude of prayer. They were both in high relief and skilfully executed. To the left of the figure first described was an inscription, consisting of thirty-six lines, in a cuneiform character resembling that in one of the columns of the trilingual tablets of Hamadan, and known as the Median or Susianian. I copied it with some difficulty, being constantly interrupted by Mulla Chiragh and his men, who suspiciously watched my proceedings, convinced that I was learning from it the site of some concealed treasure which of right belonged to them as occupiers of the country.[5]

A similar inscription had once existed near the second figure, but it had been entirely effaced by water percolating through the rock. On the dresses of both figures I could also trace remains of cuneiform inscriptions, but so much obliterated that I was unable to copy them.

[5] This inscription is included in the first volume of cuneiform inscriptions published for the trustees of the British Museum. According to Professor Sayce, it relates to the restoration of certain temples, and the carving of the sculptures and inscriptions in the Shikefti-Salman by a king whose name he reads Takhi-hi-Kutur.

On the opposite side of the cave, high up on the rocks, were two tablets which I had no means of reaching. One of them contained a group of five figures, two of which were about half the size of the others. They appeared from their postures and priestly costumes to be engaged in some religious ceremony and in prayer. In the second tablet was a similar group of three figures. From below I could see that there were also remains of cuneiform inscriptions on these tablets, but they were out of my reach, and I could not copy them.[6]

At the further end of the gorge were the ruins of an edifice of dressed stone, which may have been a fire-temple or altar. At its entrance I found some remains of buildings which appeared to be of the Sassanian period. The sculptures in the gorge are of a much earlier epoch, probably of the eighth or seventh century B.C.

The cave in which I found these sculptures and inscriptions is known to the Bakhtiyari as the Shikefti-Salman, or cave of Salman, the 'lala,' or tutor of Ali, the son-in-law and successor of the Prophet Mohammed. He is believed by the Lurs of the sect of the Ali-Ilahi to have been one of the

---

[6] For a fuller description of these sculptures see my paper in vol. xvi. of the *Journal of the Royal Geographical Society*, 1846, part i., pp. 78, 79.

many incarnations of the Deity, and to have been buried in this place, which is accordingly held in great veneration. The cave is said to extend far into the bowels of the earth, having an outlet, some of the Bakhtiyari assured me, at Shiraz!

Whilst I was copying the inscriptions and making a hasty sketch of the sculptures, I was subjected to every kind of annoyance and interruption by Mulla Chiragh and his companions. He declared that I had learnt from them where the treasure of which I was in search was buried, and insisted that I should at once disclose the place to him. A hole in the face of the rock above the tablets especially excited his curiosity, and he refused to be satisfied with my repeated assurances that I was ignorant of the contents of the chamber to which he seemed convinced it must lead. He became at last so menacing that I had to draw my long knife or dagger, the only arm which I carried, and to prepare to defend myself as I best could. Seeing that I was determined to resist, he seized my saddle-bags, opened them, and proceeded to examine their contents. After a struggle 1 succeeded in recovering them from him. As I found it impossible to complete my examination, the mulla being determined to put every obstacle in my way, I reluctantly renounced the attempt, and remounting my horse returned, with my ruffianly companions, to the encampment, resolved to complain of them to Mulla Mohammed, and if I were unable to obtain redress or protection from him, to appeal to Mehemet Taki Khan.

I was unable, in consequence of Mulla Chiragh's interference, to explore the plain near the sculptures for ruins. A few mounds a little to the east of the gorge appeared to mark the site of an ancient city—not improbably of the same epoch as the tablets—but in my hasty examination I could find no remains above ground. Some of them may cover the ruins of buildings, and there can be little doubt that a fertile plain situated like that of Mal-Emir must at one time have been thickly inhabited. There are traditions, indeed, amongst the Bakhtiyari that a magnificent palace of enormous extent once existed in it, in which the king of the land kept his treasures—hence the name, 'the treasure (mal) of the emir or prince.' But I could trace no remains of it, and the absence of dressed stone seemed to show that no such building had ever existed there.

I revisited Mal-Emir some months afterwards in order to complete and verify my copies of the inscriptions. At that time Mulla Mohammed and his tribe had moved to their pastures in the mountains, and the plain was deserted. I found myself consequently alone, and able to examine the sculptures and inscriptions without annoyance or interruption. On this occasion I succeeded in climbing to the upper tablets, but it was not so easy to descend again. I could not for a long time find a way to do so. My horse was picketed in the plain below, and I felt pretty certain that if a stray Bakhtiyari happened to pass that way he would take possession of it, and very probably, suspecting that I was deciphering the inscriptions for some mischievous purpose or for the discovery of hidden treasures, would have a shot at me. In such case I offered a capital mark, and should have been brought down to the foot of the rock without further trouble. However, fortunately I was not molested, and after remaining for a considerable time in a very unpleasant position, I divested myself of part of my clothing, and succeeded in releasing myself from it by a desperate effort, and at the risk of breaking my neck.

My complaints to Mulla Mohammed against his brother for his insolent conduct to me were of no avail. I therefore announced my determination to leave his encampment on the following morning, as I was not free to examine, as I had intended, the rest of the plain of Mal-Emir, and to proceed at once to Sûsan. The mulla then tried to persuade me to start upon my journey in the middle of the night, as the distance to Sûsan, he said, was great, and the path over the mountains so very difficult that my horse would require rest more than once on the way. Suspecting that he had some evil design in giving me this advice, I declined to follow it. He then insisted upon seeing all that I had about me, and taking possession of my saddle-bags carefully scrutinised their contents. My watch and compass particularly attracted his attention. He had never seen anything of the kind before, and made me explain over and over again their use. The compass appeared to surprise and interest him most, as he discovered that it would enable him to find the 'kibla,' or the direction of the holy city of Mecca, to which Mohammedans are required to turn when in prayer. After completing his examination of myself and of my effects, he and his followers engaged in a long and noisy squabble with the men of a neighbouring encampment, whom they accused of having stolen some of their donkeys. The dispute, which led to a violent altercation, nearly ended in bloodshed.

I had evidently fallen amongst a set of robbers and ruffians, and I had reason to remember the warning given to me by Mehemet Taki Khan's wife. There was evidently no little risk of an attempt being made upon my life. The anxiety that I could not but feel, and my doubts as to whether I should return to Kala Tul or persist in my attempt to reach Sûsan, kept me from sleeping. Mulla Chiragh, the man who had accompanied me to the rock-cut tablets, roused me about midnight, and urged me to take my departure at once. But I positively refused to leave before daylight. Although this persistent attempt to induce me to travel in the dark was calculated to excite suspicion as to the intentions of my host, I determined to pursue my journey at all hazards, and as soon as the day dawned I mounted my horse and left the inhospitable encampment. Mulla Mohammed sent with me a man who had, he said, undertaken to be my guide to the tents of Mulla Feraj, the chief of Sûsan, for whom I also had a letter from Au Baba Khan. I crossed the plain of Mal-Emir, which is only about five and a half miles in breadth and about twelve miles in length. It is ill supplied with water, and that which is found in it is considered heavy and unwholesome. Consequently it is only inhabited by Mulla Mohammed's tribe in the winter months. At other times it is deserted. During the rainy season, when the torrents descend from the mountains by which it is surrounded, the greater part of it becomes a marsh, there being no sufficient outlet for the water.

I soon reached the foot of the mountains which divide the plain of Mal-Emir from the valley of Sûsan. Through a gap in this lofty serrated ridge passed the track which I was to follow. I began the ascent to it by a gentle but very stony path. After about an hour I found myself in a narrow gorge, in the bed of a torrent then dry. I had scarcely entered it when a man suddenly appeared on the edge of a rock above me, threatening to hurl large stones upon me. The guide, who was leading my horse, from which, owing to the rough ground, I had been compelled to dismount, was a good deal in advance of me, and was evidently in league with my assailant, as he began to help himself to the contents of my saddle-bags. As I had left my double-barrelled gun at Kala Tul I was without other means of defence than my long dagger. I retreated into the bed of the torrent, and placed myself beyond the reach of the stones with which I was menaced, between two huge boulders, prepared to make the best resistance I was able, and to sell my life as dearly as possible. One or two other men soon appeared. They were without their matchlocks, but were armed with swords, which they drew and flourished in my face. Resistance would have been hopeless, and after some parley with the robbers I was compelled to deliver up my watch and my compass, and a few silver coins that I had with me. As they asked for the watch and compass and knew where I kept them, it was evident that they had been sent to rob me of them by my host of the previous night. Mulla Mohammed had apparently satisfied himself, by the examination that he had made of the contents of my saddle-bags, that there was nothing else in them worth having.

I considered myself fortunate in having escaped with my life, as my assailants would have had no scruple whatever in murdering me had they thought it necessary to do so to obtain my property, or from any other motive. I was, of course, much concerned at the loss of my watch and my compass, without which I could not make the observations required for mapping, however roughly, the country through which I passed. But I was not without hopes that, on my return to Kala Tul, Mehemet Taki Khan, whose guest and under whose protection I was, would take measures to have them restored to me.

I toiled up the difficult mountain track, following my guide, who still preceded me leading my horse, until we reached the summit of the pass, when he stopped and refused to proceed unless I gave him two tomans (about 1*l*.). Even if I had been disposed to yield to this outrageous demand I should have been unable to do so, as I had been deprived of all my money by the robbers who had taken my watch and compass. As he found that he could get nothing from me, he turned back, leaving me ignorant of the way, and night advancing—no pleasant situation in which to find oneself in these wild and lonely mountains, where every man I was likely to meet would probably be a robber, if not something worse, and where my life would have been at the mercy of the first man who chose to take it.

There was, however, nothing to be done but to continue in the track which I had hitherto followed, upon the chance of coming to habitations of some kind in the valley below, to which it appeared to lead. The mountains, which had been hitherto bare and treeless, were on the opposite side thickly wooded with oak. From the summit of the pass I looked down upon a valley through which ran the river Karun. The tents and huts of the tribe encamped at Sûsan were visible, to the north, in the distance. Entering a dense forest, I descended rapidly by a very steep and difficult path, leading my horse after me. When I had almost attained the foot of the pass my guide rejoined me. We soon afterwards came upon a few poor Iliyât families, some in black tents and others bivouacking under the oak trees. They gave me some curds, sour milk, and bread, baked as usual in very thin cakes on convex iron plates. Leaving this small encampment we descended to the river and rode along its banks. I could trace here and there the remains of an ancient paved road, and the ruins of buildings and foundations of walls. The valley was in places very narrow, with precipitous rocks overhanging the river, and we had some difficulty in making our way along it.

The ruins of Sûsan, of which I was in search, were on the opposite bank of the Karun. I had been told that I should find a raft upon which I could cross the river, but there was none. Every now and then a man would arrive with his sheepskin, blow it up, and paddle himself upon it to the opposite bank. Others were floating in the same fashion down the stream. No one, however, seemed disposed to help me. My guide shouted to some people encamped on the further side, but either his voice was drowned by the noise of the rushing water, or those whom he hailed would give no heed to him. After I had waited for some time in the hope of finding means of crossing, a fakir belonging to the tomb of Daniel promised to inform Mulla Feraj, the chief of Sûsan, that I was the bearer of a letter to him from Au Baba Khan. A raft, he assured me, would then be prepared for me in the morning.

As it was by this time nearly sunset, I made my way to a small encampment about a mile distant from the river, where I fortunately found a man who had been at Tabreez, where he had seen Europeans. He insisted upon my accepting his hospitality for the night. Protesting himself my brother, he gave me all the delicacies his tent could afford for supper.

My guide again deserted me, having made off in the night. I went down to the river bank, but there was no raft ready for me as the fakir had promised there would be. The Bakhtiyari, who were floating down or crossing backwards and forwards, were more disposed to make merry at my expense than to assist me. Determined not to be baffled, I resolved to swim my horse over the stream, and taking off part of my clothes I rode into it. It was shallower than I expected, the water only reaching halfway up my saddle, but it was so swift and strong that I had much difficulty in preventing my horse and myself from being carried away by it. However, I succeeded in gaining the opposite bank without accident.

The tent of Mulla Feraj was not far distant from the spot where I had landed. I rode directly to it. The chief received me civilly, bade me welcome, and directed a booth to be erected for my special use, which was speedily done, as there was an abundance of trees and bushes near. The mulla and his people were not less savage in appearance than the men of Mal-Emir, and their sinister looks were not reassuring. As soon as I was established in my hut the chief brought his mother, an aged, wrinkled old woman, and some of the elders of his tribe to see me. I had to undergo a rigorous and searching cross-examination. I was the first Frank that had been seen in the mountains. Was I Christian, and consequently unclean? They had heard of Georgians and Armenians, was I either the one or the other? What was the object of my journey? Had I seen in my books that a treasure was concealed at Sûsan, and did I know and could I point out to them the place where it was buried? Were the Feringhi about to take possession of the country? and innumerable questions of the kind. The prevailing opinion seemed to be that I was either a kind of magician, to whom the jinns had given the power of finding buried gold, or a secret agent sent to spy out the land. I endeavoured to allay their suspicions by saying that I was a pilgrim who had come from afar to visit the tomb of the prophet Daniel, which was known to exist in their valley, and the renown of which had reached my country. I expressed a wish to go there at once, and asked for a guide to conduct me to the sacred spot. Several men volunteered to accompany me, evidently in the hope of sharing in the buried treasure for which they were still convinced I was seeking.

I rode through extensive rice-fields, crossed an ancient bed of the river now dry, and came to a number of natural mounds, one of which had been scarped, and had apparently at some former period been surrounded by a ditch. My guides pointed it out to me as the 'kala,' or castle. On the summit there were some remains of buildings. The so-called tomb of Daniel was not far distant, at the foot of the mountains which bound the valley of Sûsan to the north. I found it to be a modern building, standing upon a small stream, containing two rooms—one open to the sky—and surrounded by a small grove of trees; but there was no reservoir with sacred fish, as described to Sir Henry Rawlinson, nor could I hear of any such fish being preserved elsewhere in the neighbourhood. Nor was the tomb of white marble, as his informant had stated, but a mean mud-built building, such as are constantly seen in the country over the tomb of some local saint. This was a fresh instance of Oriental exaggeration, proving how little trust can be placed in descriptions given by Easterns of things and places, not only of which they have heard, but which they may have seen. I was greatly disappointed, and was almost inclined to regret that I was exposing myself to so much danger and suffering on a fruitless expedition.

However, the spot is held very sacred by the Bakhtiyari, and the tradition that Daniel was buried there may be of very ancient origin. There is no doubt that throughout the mountains of Luristan the tomb of the prophet is believed to be covered by the 'Imaum-Zadeh,' or shrine, I have described. That the place and valley should be known as Sûsan, or Shushan, may add some weight to the tradition. A half-crazy dervish who had followed me to the place invited me to enter and worship at the tomb. I declined to do so, as I suspected that when he found that I was a Kafir, and had consequently polluted the sacred spot, he would denounce me to the ignorant and fanatical crowd by which I was surrounded, and the consequences might have been fatal. He could not understand why I hesitated to accept his invitation; but when one of the bystanders informed him that I was a Feringhi, and consequently an infidel and unclean, he seized a gun, and, pointing it at me, threatened to shoot me unless I repeated at once the formula of the Mohammedan profession of faith, 'There is no god but God, and Mohammed is His prophet!' Fortunately, before he could execute his menace he was disarmed. But a violent altercation took place as to the manner in which I should be treated. As a crowd of savage-looking armed men had now collected round me—men who would have made very light of cutting an unbeliever's throat — I thought it prudent to return at once to Mulla Feraj's tent. On reaching it I found that my saddle-bags, which I had left under his care, had been almost emptied of the few things that had remained in them after the repeated robberies to which since leaving Kala Tul I had been exposed. But these included my books and maps, which were precious to me. I complained loudly to the mulla of this violation of the laws of hospitality. He professed to be ashamed at what had occurred, and gave orders for the discovery of the thief, who was probably himself. My little property was ultimately restored to me.

The whole of the next day I was under much apprehension of ill-treatment on the part of these lawless savages. The brother of Mulla Feraj was in favour of compelling me to leave the encampment at once; but the mulla appeared to have some consideration for the letter of Au Khan Baba. The discussion of the previous afternoon as to my motives for coming to Sûsan, and as to the manner in which an infidel ought to be treated, were renewed. The reasons which were assigned for my visit would have been as amusing as they were ridiculous had they not been seriously entertained, and had they not endangered my life. According to one man I was employed by the Shah to examine the country with a view to its conquest and occupation. Another gravely asserted that I was the brother of the King of England, who was already at Baghdad on his way to take possession of the mountains of the Bakhtiyari. According to a third, my forefathers had buried a great hoard of gold in a spot which was described in the books I had brought with me. One arrogant fellow, who pretended to be better informed than the rest, declared positively that there had been four treasures concealed at Sûsan, and even went so far as to describe the nature of each.

But I was still resolved, now that I had reached Sûsan, to examine the remains which were reported to exist there. I therefore asked to be shown the ruins of the bridge, and of the 'mesjid,' or temple, which Sir Henry Rawlinson had been informed were to be seen there. This request further increased the astonishment and suspicion of the mulla and his followers, who were utterly at a loss to account for the knowledge of their valley which my questions displayed. I endeavoured to explain to them that some years before an Englishman, interested in the ancient history of their country, had learnt from one of their own people that remains of the city which was once its capital still existed, and that, like some learned Musulman travellers, whose works were known to all true believers, he had written an account of what he had heard about them. I had read that account, and was desirous of gratifying my curiosity by visiting the ruins. They admitted that the bridge and the mesjid were near at hand, but when I expressed a wish to see them they seemed disposed to prevent me from doing so. However, I mounted my horse, and Mulla Feraj, seeing that I was determined to have my way, ordered one of his attendants to accompany me. I had not proceeded far when several armed men joined us. I perceived that even their matchlocks were lighted. They were no doubt ready to fight for their share of the treasures which they were convinced I was about to discover. Being unarmed, I was unable to resist any violence that might be offered to me. I deemed it therefore best to assume an indifferent and unconcerned air, as. if I entertained no suspicion of my unwelcome companions.

After crossing numerous swampy rice-fields we came to the Karun, and continued along its banks until we reached a narrow gorge in the mountains, through which the river issues into the valley of Sûsan. About a mile within this gorge, in a small open space, I found the ruins of what was called the mesjid, or temple. There was nothing above ground to show that an edifice of any importance had ever stood there—no columns nor dressed stones, not even a mound, only some rough masonry, apparently the foundations of a building of the Sassanian period. These remains were however known to the Lurs as the Mesjidi-Suleiman, the temple of Solomon. At a short distance beyond them were the ruins of a bridge, of which four massive buttresses still resisted the force of the torrent. The river must have been crossed at a considerable height above the level of the stream by a single arch of great span, which was connected with the sides of the ravine by two smaller arches. I could trace on both banks an ancient paved causeway, a continuation, no doubt, of the road that I had seen in the valley of Sûsan. It was known as the 'Jeddai-Atabeg,' or road of the Atabegs, to whom its construction was attributed by the Lurs; but it was evidently a much more ancient work, possibly of the time of the Kayanian kings, and the remains of one of the great highways which in the time of Darius led from the plains of Susiana to the highlands of Persia and to Persepolis. I traced it subsequently in many places between Mal-Emir and Shuster.

The bridge had been partly built of large roughly hewn stones and partly of kiln-burnt bricks, united by tenacious cement. There was nothing that could give me a clue to the date of its construction, and as far as I could ascertain by the hasty observations I was able to make, there were no inscriptions carved on the neighbouring rocks. But it was evident that it was a very ancient structure, not later than the Sassanian period, and probably very much earlier. Beyond this bridge, higher up the gorge, the river—here an angry, foaming torrent—was shut in by precipitous rocks, and the road which was once carried along them having been destroyed, there was no means of proceeding farther. The place was called by the Bakhtiyari 'Pâyi-rah,' *i.e.* foot of the road.

Before returning to the tents of Mulla Feraj I followed the valley of Sûsan to some distance, passing on my way a few artificial mounds of no great size, and occasionally the foundations of ancient buildings, which were sufficient to show that at one time a city of some importance might have existed here. Black tents were scattered among them, the owners of which, with more hospitality than the behaviour of the mulla's followers had led me to expect, invited me to dismount and to eat bread with them. One old man, who protested that he was above one hundred years of age, and that he had lived in the reigns of six shahs, declared that he had never seen a Feringhi before in Sûsan, and had never heard of one having been there.

I was much struck by the attention I received from a man who told me that he had served in a Bakhtiyari regiment of regular troops raised by Abbas Mirza, in which were English officers. He so pressed me to breakfast with him that I could not refuse. To my surprise he dipped his hand into the same dish with me—a thing that no Persian or Bakhtiyari had hitherto done—observing to those around him that he had seen Englishmen eat with the Prince and with other great personages, and that they were not like other Kafirs, who were unclean. He invited me to pass the night in his tent, and when he found that I was unable to do so, he filled my saddle-bags with pomegranates and dried fruit.

In the evening, on my return to Mulla Feraj's encampment, the endless discussions about my object in coming to Sûsan were again revived. Fortunately, the attention of my hosts was soon diverted from this subject by two musicians who arrived at the tents and played on the drum and a kind of oboe. A crowd of men and women gathered round them, their savage swarthy faces lighted up in a ghastly manner by a blazing fire. They seemed to be greatly excited, as those mountaineers usually are, by the wild music, expressing their feelings, according to the melody, by loud deep-drawn sighs or by warlike shouts.

It was only on the following morning, when the discussion and cross-examination about the object of my visit were resumed, that I heard for the first time that there was an inscription carved on the rocks near the Pâyi-rah. I was told that it was in the writing of the Feringhi, and only three or four lines in length, and that it was a 'telesm,' or talisman, which indicated the spot where the treasure of which I was in search was buried. Some of my tormentors were of opinion that I ought to be taken to it, and compelled by force to disclose the secret. Others insisted that I should not be allowed to see it. As they began to hold very menacing language, and seemed disposed to proceed to acts of violence, I thought it more prudent to give up any further attempt to explore the valley of Sûsan. I had learnt that the mother of Mulla Mohammed, the chief in whose tent I had stopped in Mal-Emir, was on a visit to Mulla Feraj, to whom she was related, and was to leave for her home in the course of the morning, accompanied by her female attendants and by some armed men. I determined to join the party, as I was not without apprehension that if I returned alone I might be robbed on my way by some of the mulla's people, or that, fearing I might complain to Mehemet Taki Khan of his treatment of me, he might even cause me to be murdered. His brothers had never ceased asking me for almost everything that I possessed,—even to some of my clothes—and I was compelled to give them the bridle and the greater part of the trappings of my horse. Some of the contents of my saddle-bags had, moreover, been stolen. Even the shoes of my horse had been taken off.

I was not sorry when I turned my back upon the importunate and inhospitable Mulla Feraj and his tribe. I joined the small caravan going to Mal-Emir which had assembled near his encampment. At a short distance from it I passed, near the river bank, foundations of buildings, remains of ancient walls, and other ruins, which were known to the Bakhtiyari as Mali-Virun. My companions pointed out to me what their imagination led them to describe as streets, bazars, palaces, and castles. I could, however, trace what appeared to be a triple wall once protecting the city on the northern side. The masonry of these remains was of rounded stones from the river, united by a very tenacious mortar, characteristic of the Sassanian period.

Mali-Virun and the other ruins I have described were all that I could find in the valley of Sûsan. They were far less extensive and important than I had been led to expect. I was unable to discover any architectural remains, columns, slabs with inscriptions, or stones, or bricks with distinctive marks, such as are found among ruins of the Kayanian period, nor even any dressed stones, except those employed in the construction of the bridge at Pâyi-rah. But the traditions of the Lur tribes point to the place as the site of a very great and ancient city, and to the tomb of Daniel as the true burial-place of the prophet—the 'greater Daniel' as they term him. That on the river Shaour, in the plains of Susiana, which is on the site of the ancient Susa, they assign to Daniel Askar, or the 'lesser Daniel,' and hold it in less reverence. The valley of Sûsan, from its natural strength and from its fertility, might well have been chosen for the site of a considerable city. It now produces, although only rudely cultivated by the Bakhtiyari nomads, an abundance of rice, corn, and barley, and figs, pomegranates, and other fruit. It is well irrigated by the Karun, and by the numerous streams which descend from the surrounding mountains, and its water is renowned for its purity and wholesomeness. This river is broad, and navigable for laden rafts, from the place whence it issues from the mountain gorge.

We had to cross the Karun before reaching the

hills which separated us from Mal-Emir. A small raft made of a few inflated skins and some bundles of reeds had been provided for the women, but its owner refused to take me across on it unless I paid him. This I was unable to do, as I had been robbed of all my money. After a good deal of wrangling, and on the intervention of Mulla Mohammed's mother, the matter was arranged by my giving the man medicine for his eyes. We were obliged to pass the night under some trees, where we found a party of Bakhtiyari resting on their way to their winter pastures in the low country. The women of our party were received by those of the encampment with wailings, as one of them had recently lost her husband. They all seated themselves round a fire, tearing their hair, beating their breasts, and howling, like the mourners at an Irish wake. After about an hour, they set to work to cook the dinner. As the principal guest was the mother of a chief, a sheep was slain for her. These wanderers seemed, however, to be very poor, and were destitute of almost everything, eating bread made of acorns.

We left our forest encampment in the middle of the night, and having crossed the mountains before dawn, reached Mulla Mohammed's tents early in the day. Mulla Chiragh, his brother, whom I accused of having been the instigator of the robbery committed upon me, and of possessing my watch and compass, absolutely denied all knowledge of the matter. But on my arrival on the following morning at Kala Tul, I denounced him to Au Baba Khan as having treated with contempt the letter which he had given me, and as having brought disgrace upon the Bakhtiyari name by robbing and ill-treating one who had been his guest and had eaten his bread. The chief heaped foul epithets upon him and his father and mother, and sent off a man at once with orders to bring back my watch and compass without fail. They were both recovered and restored to me, and fortunately without having suffered material damage.

Another expedition I made whilst at Kala Tul was to the ruins of Manjanik, of which also Sir Henry Rawlinson had received a most exaggerated account from his Lur informants. To reach them I had to cross the small plain of Baghi-Malek, 'the King's Garden,' through which runs a clear mountain stream called the Abi-Zard, upon which they stand. The plain, notwithstanding its attractive name, is barren and stony, and thinly wooded with the 'konar,' or jujube tree, which bears a sweet fruit something like a small date. The Bakhtiyari say that it was here that Abraham was cast from afar into a fiery furnace by Nimrod, with a 'manjanik,' or 'mangonel,' a kind of military engine like the catapult—the heat of the fire being so great that none could approach it. This legend is, I believe, founded upon a Rabbinical tradition which has found its way into the Koran, where Abraham's escape from 'Ur of the Chaldees' has been translated into 'the fire of the Chaldees.' According to some commentators on the Koran, a Persian Kurd, named Heyyun, counselled Nimrod to commit this outrage. Sir Henry Rawlinson had been led to believe that a great mound and extensive ruins existed on this spot, and conjectured that it might be the site of the ancient city of Seleucia, or Elymais, mentioned by Pliny. But the only remains that I could discover there were some ruined buildings with vaulted chambers, constructed of rounded stones from the torrent united with cement, similar to those which are found in all parts of the mountains, and which are probably of the Sassanian period. The walls of some of these rooms were covered with fine stucco, and decorated with mouldings and ornaments still well preserved, and resembling those in the palace of Ctesiphon and in buildings in various parts of Persia of the same period. A few ruins on a natural mound on the bank of the stream appeared to be those of a small castle. Among them was the tomb of a local saint venerated by the Bakhtiyari. Near the mound the Abi-Zard had been spanned by a bridge, three piers of which still remained, and I discovered what appeared to be the basement of a tower or minaret. The large village of Manjanik was surrounded by rice-grounds and gardens, chiefly containing pomegranate trees, which abound in this part of the country.

The huts of the chief, or Aga, and of his followers, were constructed of boughs of trees and reeds. Some of them were larger than any I had previously seen, and lofty and spacious, the roof being supported by numerous poles, to which were hung skins containing butter and curds, and various utensils for cooking and other purposes. There were also some black tents, and many enclosures to protect the sheep and cattle from wolves and other beasts of prey during the night.

Early in November I had a severe attack of ague. Kala Tul is, at this time of the year, very unhealthy, and there was scarcely a family of which most of, if not all, the members were suffering from fever. I was so ill that Mehemet Taki Khan's wife proposed that I should accompany her and two of the children—one of them, my little patient, Hussein Kuli—for change of air to Boulabas (a corruption of Abu'l-Abbas), a village on the Abi-Zard river, and in a small but highly cultivated valley filled with fruit trees. Khatun-jan Khanum was received with great respect by the inhabitants, who came out to meet her. The best house, which was sufficiently spacious and of stone, but in a ruinous condition, was placed at her disposal. With rest and by the help of quinine I soon recovered from my illness. The village, which contained about three hundred houses, was built upon the site of an ancient town, a few remains of which still existed, and were known by the name of 'Kala Giaour,' or 'Kala Gebr'—the castle of the infidels. At a short distance from it was an 'Imaum-Zadeh,' or shrine, sacred to Solomon, who, according to local tradition, visited the spot, which is called Rawad. The river, which has its source among the snows of Mungasht, issues from the mountains not far from the village, through a grand gorge, wooded by magnificent trees. I remained some days at Abu'l-Abbas, nursed with the kindest care by Khatun-jan Khanum. As soon as I felt able to resume my wanderings I determined to pay a visit to Shefi'a Khan, whom I had accompanied from Isfahan. I was furnished by Khatun-jan with a letter to Zacchi Aga, the chief of a small tribe inhabiting the plain of Baghi-Malek. He was directed to send a guide with me to the tents of my friend, who was encamped near a spot where I had been told there were ancient inscriptions. I had scarcely reached Zacchi Aga's tents when I was seized with a fresh fit of ague.

Early next morning, however, I was able to continue my journey with a small caravan of men on foot with donkeys, laden with rice, going to Shuster. The owners assured me that their route lay through Shefi'a Khan's encampment. We crossed during the day one or two ranges of low hills, and an uncultivated, undulating country without inhabitants. In the distance, to the northwest, rose a barren mountain, called Ausemari. The head of the caravan informed me in the afternoon that he had learnt that Shefi'a Khan had recently moved his tents from the plain to the foot of this mountain, and that it would be out of his way to go to them. Showing me the direction in which he believed them to be, he advised me to strike across the country to them. It was already late in the afternoon, as our progress with the laden donkeys had been very slow. I was told, however, that I should be able to reach the tents before the sun went down. As I put little trust in this assurance, I should have returned to Manjanik had I been able, but there would not have been time for me to reach the village before late at night, and it was far from safe to travel after dark. I therefore left the caravan and rode off in the direction of Ausemari.

I was now alone. Night was coming on. The country was dangerous, and I might fall in with a solitary robber or with horsemen on a raid. Lions, too, were not rare in those plains. There was no encampment to be seen in the distance, nor did I meet a human being. I began to fear that I should have to pass the night in this desert, without food for myself or my horse, and without even water. The sun had just disappeared when I perceived two men on foot. I urged on my tired horse and soon overtook them. Fortunately, they proved to be inoffensive people, and offered to conduct me to some tents which were near. I should probably not have discovered the small encampment had I not been guided by them to it, as it was carefully concealed in a deep gully, in order to escape observation from marauders. The 'Ket-Khudâ,' who was the chief of a few Bakhtiyari families, received me hospitably, and at once found provender for my horse and supper for myself.

Although I had been well received by the chief, I was by no means persuaded that he or some of his people might not have designs upon the little property I had with me. As he declared that the encampment of Shefi'a Khan was still far off among the hills, and declined to give me a guide to it, I made up my mind that, after following for a short distance the track which was pointed out to me, I would turn back and make the best of my way to Abu'l-Abbas. My suspicions had been aroused by seeing two men leave the tents early in the morning. They were confirmed when, on leaving the track and going in the opposite direction, I saw these men running after me and making signals for me to stop. I put my horse to a canter and was soon out of their sight.

I had not ridden far when I was seized with so severe an attack of ague that I had to dismount, and, hiding myself in a gully, to lie down on the ground, with the bridle of my horse fastened to my wrist. I remained delirious for two or three hours, as was usual with me. Fortunately I was not discovered. After this stage of the fever had passed I felt able to continue my journey, and reached Abu'l-Abbas soon after sunset.

Shortly after, on my return to Kala Tul, Shefi'a Khan himself came to the castle. When he had finished his business he invited me to return with him to his encampment. He and his family and followers were living in regular Iliyât fashion, in large black tents pitched in a valley in the rocky and treeless mountain of Ausemari. The country through which we rode to reach them was well watered and fertile. It was a favourite winter camping-ground of the Bakhtiyari, and their tents were to be seen in every direction.

I found some of my travelling companions on our journey from Isfahan among the followers of the Khan, and received from them a very friendly welcome. During the few days I spent with him, he gave me much interesting information about the Bakhtiyari country and tribes, which will be found in my paper upon Khuzistan, in the 'Journal of the Royal Geographical Society.'[7] He was very intelligent, could describe with sufficient accuracy, and was always ready to communicate what he knew, not having those absurd suspicions as to my motives for asking questions and for visiting his country, which had been the source of so much annoyance and danger to me. As he was Mehemet Taki Khan's principal adviser and vizir, and was employed in administering the affairs of the tribes, and in apportioning and collecting their respective contributions to the tribute payable to the Persian Government, he was better acquainted with all that concerned the Bakhtiyari and their history, the number of their families, that of the horsemen they could send to war, and other matters, than any one I knew. I found that the information he gave me could be relied on as trustworthy.

The winter had now set in, and whilst I was with Shefi'a Khan there were constant heavy rains,

[7] Vol. xvi.

with thunder, lightning, and high winds. The 'lamerdoun,' or guest tent, offered but little protection when these storms broke over us, and I was frequently drenched to the skin during the night. On one such occasion a pack of wolves made a descent in the darkness upon the sheep, and breaking through the tents, carried off nine of them. The screams of the women, the cries of the men, and the barking of the dogs—the thunder rolling in awful peals and the lightning flashing with the most dazzling brightness—added to the terrors of the night. Tents were blown down. Torrents from the hills swept into the plains carrying everything before them. We had to seek for refuge behind rocks and wherever we could obtain shelter. The horses, terrified, broke loose from their tethers and fled. Such a night I had never before and have never since witnessed.

The desolate hills of this part of Khuzistan abound in wild animals. In addition to wolves, which are much dreaded by the shepherds, lions, leopards, bears, lynxes, wild boar, hyenas, jackals, and other beasts of prey, and various species of wild sheep and goats, are found in great numbers in them. The Bakhtiyari chiefs delighted in the chase and were constantly engaged in it.

To kill a lion, especially in single combat, was considered a great feat, and the figure of a lion rudely carved in stone is placed by the Bakhtiyari over the graves of their warriors, to denote that they were men of valour and intrepidity. Mehemet Taki Khan was renowned for his skill and cool courage in these encounters, and other chiefs were celebrated for victories they had achieved over this ferocious and wily beast. Whilst I was living with the Bakhtiyari I was present at more than one lion hunt. One afternoon when Mehemet Taki Khan was seated at the doorway of his castle with the elders, as was his wont, a man arrived breathless and in great excitement, declaring that in crossing the plain he had met a lion in his path. The beast, he said, was preparing to spring upon him, when he conjured it in the name of Ali to spare a poor unarmed man, who had never harmed any of its kin. Thereupon, the lion being a good Musulman and a Shi'a to boot, as some lions are believed to be, turned away and disappeared among some bushes.

The man, ungrateful to the lion who had spared him so generously, offered to conduct Mehemet Taki Khan to the spot where the beast had left him. Although the chief doubted the truth of the story, some horsemen and matchlockmen on foot were assembled, and we left the castle with our guide. He led us to a kind of pit or hollow in the ground, filled with low bushes, in which, he said, the lion had concealed itself.

Mehemet Taki Khan divided his horsemen into three parties, placing one of them under his brother, Au Khan Baba. Stones were thrown and guns fired into the thicket, and other means taken to drive the animal out of it, but in vain. The Khan's suspicions that the man had been frightened by a hyena or a wolf, and had invented the story of the lion, were confirmed. Whilst we were deliberating as to returning to Kala Tul, the animal, roused by a man who had descended into the hollow, suddenly sprang towards Au Khan Baba, with whom I had placed myself. He fired with his long gun and wounded the lion, which, however, passed by him and seized a matchlockman named Mulla Ali, who in falling caught the dress of Mehemet Ali Beg, whom he dragged down. Both men were thus in the lion's power and in the most critical situation.

Mehemet Taki Khan himself jumped off his horse, and advancing towards the beast addressed it thus in a loud voice: 'O lion! these are not fit antagonists for thee. If thou desirest to meet an enemy worthy of thee, contend with me.' The animal did not appear disposed to abandon its prey, which it was holding down under its massive paws. It raised its head majestically as if defying its numerous foes. The chief approached it, and drawing the long pistol which he carried in his girdle, fired at its head. The bullet took effect, and the lion falling to the ground was quickly despatched by the guns, swords, and spears of Mehemet Taki Khan's followers.

The lion, which was pronounced to be an unusually large one and had a short black mane, was borne in triumph to the castle. Its skin was presented to me, but I was afterwards robbed of it as of other things.

Mehemet Ali Beg was seriously hurt, one of his arms being badly crushed and the flesh torn from one side of his face. The matchlock-man received one or two wounds of less consequence. During my residence in the Bakhtiyari Mountains the story of the great chief's valour and prowess, and how he had addressed the lion, formed a constant theme of conversation in the tents, and, I have no doubt, has remained a tradition amongst the tribes.

On occasions when I accompanied one of the chief's brothers on regular lion hunts we went to the banks of some stream covered with reeds and bushes, their usual haunts. Beaters were sent into the jungle and the horsemen remained outside in the plain. Generally only wild boars were driven out and were pursued and shot; sometimes a lion was disturbed, and leaving its lair bounded across the plain. It was followed by the horsemen, but was rarely overtaken and killed, unless it took refuge in the low brushwood. It seldom turned upon its pursuers when in the open, and gave no proof of the courage with which it is generally credited. But when suddenly disturbed, or surprised in its retreat, by the beaters, it attacked them with great fury, and more than once a man was killed in this manner.

The Asiatic lion appears to differ from his African fellow in courage and daring, as well as in size and strength. It will rarely attack a man unless provoked, or driven to do so by hunger, and at the sight of one usually slinks off and hides itself. In the night it will creep into an encampment and seize a bullock or a sheep or even a sleeping man. One of a party of Bakhtiyari with whom I was hunting at the foot of the hills near Shuster was thus carried off. We had to sleep on the ground in the open air. In the morning one of the men was missing, and his remains were discovered not far from the spot where we had passed the night.

The Susianian lion is, nevertheless, a formidable animal, and stories of encounters with it, and of travellers who have been attacked and devoured, form part of the staple of the evening's talk in a Lur tent. As to its strength, the Bakhtiyari allege that it can carry off a full-sized buffalo or an ox, but not a sheep, for, they say, when it bears away a buffalo it invokes the aid of Ali, but when a sheep it relies upon its own strength. Shefi'a Khan, however, attempted to explain this alleged fact to me by suggesting that whilst the lion could throw a large animal like a cow or buffalo over its back it was obliged to trail a sheep on the ground, and to abandon it when pursued.

It is more to flocks and herds that the Asiatic lion is formidable than to man. Amongst them it makes great depredations, destroying and carrying away sheep and oxen. Buffaloes, however, are said to beat it off by placing themselves back to back, and meeting their assailant with their bulky foreheads and knotted horns. Horses are much terrified at the sight or smell of a lion. Its vicinity to an encampment is soon known by the uneasiness and fear shown by the horses, who snort and rear, and struggle to break away from their tethers. The young chiefs, as I have already mentioned, accustom their steeds to the sight and smell of the animal by taking them up to a stuffed lion's skin.

Among some memoranda written at Kala Tul I find the following notes about lions and other wild beasts.

The lion abounds in the district of Ram Hormuz and on the banks of the Karun. It frequently ascends, in search of prey, to the higher valleys at the foot of the great chain of the Lur Mountains. During my residence here (Kala Tul) several have been seen in the neighbourhood, and a large lioness was killed a short time ago by a matchlock-man in the teng' (defile) of Halaugon. She measured $10\frac{1}{2}$ feet in length. Lions in this country are sometimes very bold and fierce, and are consequently much dreaded by the Iliyât. They frequently rush into the middle of an encampment, and carry off horses and other animals. I have heard many well-authenticated stories of such attacks. It is said that the buffalo does not fear a lion, and will even drive it away, whilst other animals are paralysed by fright at its approach. Therefore, in the plain of Ram Hormuz, the Bakhtiyari place male buffaloes outside the encampment as a guard. The Lurs pretend that on the approach of a lion the buffalo will summon it to retire, and if not obeyed will drive so furiously with its powerful horns at the beast, that it will be glad to take to its heels.

The Lurs divide lions into Musulmans and Kafirs (infidels). The first are of a tawny or light yellow colour, the second of a dark yellow, with black mane and black hair down the middle of the back.[8] If, they say, a man is attacked by a Musul-

[8] Probably the lighter in colour are the females, the darker the males.

man lion he must take off his cap and very humbly supplicate the animal in the name of Ali to have pity upon him. The proper formula to be used on the occasion is the following : 'Aï Gourba Ali, mun bendeh Ali am. As khana mun bigouzari. Be seri Ali '—*i.e.* 'O cat of Ali, I am the servant of Ali. Pass by my house (or family) by the head of Ali.' The lion will then generously spare the suppliant and depart. Such consideration must not, however, be expected from a Kafir lion. The Lurs firmly believe in this absurd story.

A single lion will frequently cause considerable mischief. During a period of three years one haunted the plain of Ram Hormuz. Scarcely a night passed without a human being, a horse, or a cow falling its victim. It never appeared in the same place for two days running. It cunningly evaded every attempt to destroy it. No place was secure from its attacks, and it would enter huts and tents in pursuit of its prey. It was at last killed when, in the spring, the Matamet with his army passed through the plain. During the night it had carried off a soldier whose remains were found. The beast was traced to a thicket, and a detachment of the Luristan regiment succeeded in slaying it, though not until it had severely wounded two men and had been pierced by several balls. I saw it when dead. It was unusually large, and of a very dark brown colour, in some parts of its body almost approaching to black.

Mehemet Ali Beg related to me how, on one occasion, as he was striking his tents to move up to the 'sardesirs,' or summer pastures in the mountains, a lion suddenly dashed into the midst of the women who were waiting to commence the march. Some were on horseback, others on foot. The greatest confusion and alarm prevailed. Several of the women were knocked down, but were not injured by the animal, which threw itself upon a horse. It happened to be that upon which Mehemet Ali Beg's wife was riding. He flew to her rescue, and addressing the savage beast, according to the custom of the Lurs, in some such words as 'O lion! what hast thou to do with women? Dost thou fear to face a man like me?' despatched it with a shot from his long gun.

The lion has not, I believe, been known to traverse the high chain of the Luristan Mountains into the valleys on the Persian side. In the plains of Khuzistan its usual places of concealment are the brushwood and jungle on the banks of rivers and streams and in the rice-grounds.

The Lurs and Arabs pretend that formerly an animal which they call the 'uze' was found in the jungles of the Karun and Kerkah rivers, but that it is now extinct. It had the swiftness of the greyhound, which it resembled in its limbs. It was a very ferocious beast, and was greatly dreaded. From the description they gave of it their account was either greatly exaggerated or was probably entirely fabulous.

The Bakhtiyari Mountains contain leopards of great size and fierceness. They rarely, however, attack men, but frequently carry off cattle and sheep. Their skins were occasionally brought to Kala Tul. The chiefs made saddle-clothes of them.

I have only seen one kind of bear in the Bakhtiyari Mountains. It is of a pale dirty-brown colour, and attains a considerable size. It is not much feared by the Lurs, and rarely destroys sheep or cattle. It is probably the 'Ursus Syriacus.' The Bakhtiyari have a number of strange stories and traditions connected with the bear.

Whilst I was staying with Shefi'a Khan a horseman arrived from Kala Tul urging me to return there at once, as both Mehemet Taki Khan and his brother, Kelb Ali, were seriously ill. I accordingly left the encampment and was overtaken by another terrific thunderstorm. In crossing the hills I could scarcely retain my seat on my horse, such was the violence of the wind. Impetuous torrents swept through the gullies and watercourses which three or four days before had been entirely dry. We had to make long detours to avoid them, leaving the beaten tracks and scrambling over rocks and stony ground. The waters were out in all directions, and the plain of Tul had the appearance of a lake. In crossing a swollen stream my horse was carried from under me, but succeeded in swimming to the opposite bank. Encumbered by my heavy Bakhtiyari felt outer coat, which prevented me from using my arms, I was swept down to some distance, and should inevitably have been drowned had not my guide, who had crossed safely, ran to my help.

Mehemet Taki Khan was suffering from a slight bilious attack, from which he speedily recovered. But his brother appeared to be in a hopeless state, as his fatal malady was making such rapid progress, that even the most skilful physician would have been powerless to arrest it.

CHAPTER X.

Demands upon Mehemet Taki Khan—He is declared in rebellion—Threatened invasion of his mountains—Requests me to go to Karak—The trade of Khuzistan—Leave for Karak—The Kuhghelu—Ram Hormuz—The Bahmei—Behbahan—Bender Dilum—Mirza Koma—Embark for Karak—Arrive there—Return to Kala Tul—March with Mirza Koma—Danger from Arabs—Reach the castle—Mehemet Taki Khan at Mal-Emir—Adventure with Baron de Bode—Join Mehemet Taki Khan—Effect of poetry on Bakhtiyari.

AT the end of November Mehemet Taki Khan received letters from Tehran and elsewhere which much disquieted him. The Matamet, within whose government the Bakhtiyari tribes were included, had been constantly making demands upon him for arrears of tribute. Several persons having 'berâts,' or Government orders for money, upon him had arrived at the castle. The Matamet's 'shutur-bashi,' who had accompanied Shefi'a Khan from Isfahan, had been sent to collect ten thousand tomans (about 5,000*l*.), three thousand of which were to go to Tehran as part of the revenue which was due to the Shah, three thousand were for the Matamet himself, and the remainder was to satisfy various claims made upon the royal treasury by private individuals. The usual mode of settling such claims was by giving the claimants drafts upon villages, tribal chiefs, or wealthy notables, and leaving them to get them cashed as they best could. The bearers of these documents, which were frequently sold by the original possessors with a very large discount, generally quartered themselves upon the persons upon whom they were drawn, and remained for many months— even years— until the sum for which they were given was paid. Such was the case with some 'berâtdars,' as they were called, at Kala Tul. Their presence, as may be supposed, was far from being agreeable to the chief, but it was not considered prudent to dismiss them without first satisfying them by the payment of part, if not the whole, of their claims, and they were lodged in the castle and treated as guests.

Mehemet Taki Khan had hitherto evaded the payment of the ten thousand tomans demanded by the Matamet through the shutur-bashi. He had not so large a sum at his command, and to attempt to collect it from the semi-independent tribes under his jurisdiction would have been to run the risk of causing conflicts with them, leading to bloodshed, which would have seriously weakened his influence and authority in the mountains. The

Bakhtiyari had very little ready money, and what little they had they were not very willing to part with. During the whole time that I was with them I rarely saw a gold or silver coin, except such as were worn as ornaments by the women. They had little or no trade, not sending much of the produce of their mountains and valleys for sale to the settled districts and towns of Persia. Amongst themselves it was considered disgraceful 'to sell bread,' and as the laws of hospitality are universally recognised as obligatory upon Musulmans, no one was required to pay for the food which he might consume when in a Bakhtiyari tent. They cultivated sufficient corn and rice for their immediate wants; they made their clothes and their tents out of the wool and hair produced by their flocks and herds; and the few European goods they required were usually obtained from itinerant traders who received produce in exchange for them.

To collect the sum demanded by the Matamet extreme measures would have been necessary, such as torture, without which Persians of all ranks would rarely part with their money, or the use of force in the case of a refractory tribe. To none of these measures would Mehemet Taki Khan have recourse. He, therefore, sought every kind of excuse and every means of delay to avoid the payment of the tribute and the other claims upon him.

Constant pressure was exercised upon him by official communications, but in vain. At length he received a letter from his brother, Ali Naghi Khan, who was kept at Tehran as a hostage for his loyalty and good behaviour, informing him that the Matamet had complained to the Shah that he had been in secret correspondence with the exiled princes at Baghdad, that he refused to pay his appointed tribute, had dishonoured the Government drafts upon him, and was therefore 'yaghi,' or in rebellion. His Majesty had consequently directed the Governor of Isfahan to take such measures to enforce the royal authority as he might deem necessary, and a military expedition was to be sent in the spring, as soon as the mountain passes were open, to invade and occupy the Bakhtiyari country.

The Persian Government had long been jealous of the power of Mehemet Taki Khan, who had succeeded in bringing so large a portion of the Bakhtiyari tribes under his sway, and suspected him of a design to throw off his allegiance altogether. The most exaggerated accounts of the wealth supposed to have been accumulated by the Bakhtiyari chief had also reached the Shah, who, after the fashion of Persian sovereigns, considered that the greater part, if not the whole of it, ought to be transferred to the royal coffers. Mehemet Taki Khan, rather than engage in open war with the Persians, had consented to the retention of his brother Ali Naghi at the capital as a hostage. But he writhed under their constant exactions; he deplored the tyranny and maladministration which were the cause of widespread anarchy and disorder, and were bringing the kingdom to ruin, and he despised the pusillanimous and corrupt Persian authorities. He had, moreover, much contempt for the Persian regular army, which was at that time badly armed and ill-disciplined. But, nevertheless, he still hesitated, as is usually the case with such semi-independent tribal chiefs, to declare himself in open rebellion. He sought to temporise and to ward off, if possible, an invasion of his mountains, and a conflict in which some of the tribes he had brought under his authority might be induced, by intrigues at which Persians are adepts, to join the invaders against him. The Persian and Turkish Governments, in order to maintain their rule over the warlike inhabitants of Kurdistan and Luristan, have followed the policy which has prevailed in those wild regions from time immemorial. It consists of the *divide et impera* system—setting one tribe against another, and bribing the principal chiefs with gifts, or with promises of support in their struggles one with another for the headship of their clan.

The letters from Ali Naghi Khan had consequently caused his brother great anxiety, which was increased by the reports that reached him from Isfahan that the Matamet was already making preparations by collecting a force of regular troops and artillery, to invade the Bakhtiyari Mountains as soon as the season would permit. Shefi'a Khan was hastily summoned to Kala Tul to give his advice as to the course to be pursued. I was present at some of their discussions, and was asked for my opinion, which I was very reluctant to express. It was decided, at last, on Shefi'a Khan's recommendation, that every effort should be made to come to an arrangement with the Matamet, in order to prevent war and an invasion of the mountains, and that he himself should, with this object, visit the various tribes under Mehemet Taki Khan's authority, with a view to collecting as much money as possible to satisfy the demands upon him, but that they should not be called upon to furnish their contingents of armed men, and that other measures of defence should not be taken which might furnish an excuse to the Persian Government to proclaim Mehemet Taki Khan in rebellion against the Shah.

Although the chief was ready to act upon

Shefi'a Khan's advice, he was convinced that the Matamet had resolved to invade the Bakhtiyari Mountains, and to make him a prisoner, whatever proofs he might give of his submission and of his loyalty An expedition against him, if successful, would enrich its promoter. The inhabitants of the country invaded would be robbed and plundered by the Persian officials and soldiery, until they had scarcely the shirts on their backs left to them. He would be accused of being 'yaghi,' his property would be confiscated, and if he fell into the hands of the Matamet he would probably be put to death, or at any rate be sent, with his wives and family, a prisoner to Tehran, deprived of his sight, and kept in chains for the rest of his life.

He deemed it necessary, therefore, to take some precautions to prevent the consequences he anticipated. He knew that a quarrel between the English and Persian Governments had led to the recall of the British representative from Tehran, and to a suspension of diplomatic relations between the two countries. Rumours had reached him that they were on the verge of war. These rumours were confirmed by the news which came to Kala Tul of the occupation of Karak, in the Persian Gulf, by British troops. The report that an English army, with innumerable cannons, had taken possession of this island, and was about to cross to the mainland in order to advance upon Shuster and Shiraz, had spread through Luristan and among the Arab tribes inhabiting the plains between the mountains and the Euphrates. I do not think that Mehemet Taki Khan had entirely divested himself of the suspicion that I was a British political agent entrusted with some secret mission. He probably hoped that if war were to break out between England and Persia he might avail himself of the opportunity to proclaim his independence. He had at his command many thousands of the finest and most daring horsemen and most skilful matchlock-men in Persia, and he had reason to believe that the force already at his disposal might be greatly increased should he bring about a general rising against the Shah, to be supported by English money, bayonets, and artillery. He was desirous, therefore, of communicating with the British authorities at Karak, and learning whether, in the event of war, they would be prepared to accept his assistance, and to enter into an agreement with him to protect him against the vengeance of the Shah, and to recognise him as the supreme chief in Khuzistan on the conclusion of peace. He accordingly begged me to proceed to that island in order to ascertain if possible, the intentions of the British Government, and to submit his proposals to the commander of the British forces there.

There were other reasons which induced me to accede to Mehemet Taki Khan's request to proceed to Karak. I was anxious to ascertain whether it would be possible for me to do anything to save, or prolong, the life of Au Kelb Ali, his brother, whose malady appeared to be making rapid progress. I might be able to obtain advice and medicines from some physician attached to the British force there, which would enable me, at least, to alleviate his sufferings. Mehemet Taki Khan, who had great affection for his brother, and the young chief's wife, who was nursing him with the tenderest care, were earnest in their entreaties that I should do so, and the Khan seemed to attach as much importance to my journey to Karak on this account as he did to its political object.

Again, it was now many months that I had been without news from England. I should, moreover, enjoy for a short time the society of my countrymen, of which I had now been long deprived.

Mehemet Taki Khan was a man of broad and enlightened views, notwithstanding his want of anything like education, and although he was only the chief of wild mountain tribes. Of an evening, when sitting together in the enderun, he had often spoken to me of his desire to put an end to the lawless habits of the Bakhtiyari, to introduce order and peace into his country, and to develop its resources. I pointed out to him how this could be best done by encouraging trade and entering into communication with civilised nations. I showed him that the province of Khuzistan produced many things, such as cotton and indigo, that were highly prized in Europe, and that its carpets and other manufactures were equally esteemed, and that British and other merchants might be encouraged to establish a trade in them, which would have the effect of inducing his people to engage in peaceful pursuits, and would enable them in return to obtain from England and elsewhere many necessaries and luxuries of which they were in need, and which would contribute greatly to their well-being. He informed me that, wishing to open such a trade between his mountains and India, he had entrusted a Christian with a cargo of the produce of the country, which was shipped in a native vessel at Muhammera, at the mouth of the Karun. The ship, with its contents, was lost in the Persian Gulf on its way to Bombay.

He readily entered into my views, and authorised me to inform the British authorities at Karak that he was prepared to make roads through that part of the country which was under his authority and control, and which at that time extended to the plains inhabited by Arab tribes almost to the

Shat-el-Arab, or Euphrates, and to the upper part of the Persian Gulf. He begged me to endeavour to induce British merchants to trade with his people, promising them complete security for themselves, their agents, and their property. During the time that I had been with him I had made inquiries as to the produce of the country which might be profitably exported. It consisted chiefly of indigo, cotton-wool, goats'-hair, gall-nuts, wax, the sweetmeat called 'gazu' or 'gazenjubîn,' rice, and various kinds of cereals. I ascertained their prices, and the cost of carriage, with other particulars which might prove useful to any Englishman who might be willing to attempt to open a trade with the province of Khuzistan. But I did not expect to find any such person at Karak, which was merely a military station. I, however, looked forward to doing so should I return to Baghdad, where more than one enterprising British merchant was then established.

Mehemet Taki Khan was about to send Mehemet Ali Beg—he who had the adventure with the lion—upon a political mission to Mirza Koma, the chief of Behbahan,[1] a town in the low country between the great range and the Persian Gulf. I was to accompany him with a letter for the Mirza,

[1] These names were so pronounced by the Bakhtiyari. They should properly be written Kumo and Bihbihân.

requesting him to send me with a guide, or an escort, if necessary, to Bender Dilum, a small port on the Persian Gulf, where I should be able to find an Arab sailing-boat to take me to Karak. Our departure was delayed for several days until the mulla, who had to be consulted before the journey was commenced, had obtained a favourable omen. At length, on December 8, that day having been pronounced propitious, I took leave of Mehemet Taki Khan and his wife and children, who showed real grief at my departure, and, mounting my horse towards evening, rode off with Mehemet Ali Beg, promising to return as soon as possible.

As we had left the castle late in the day we could not proceed beyond the plain of Baghi-Malek,[2] where we stopped at an encampment for the night. We started before sunrise on the following morning, as we had a long day's journey before us. We passed through the ruins of Manjanik, which I had already visited, and crossing a steep and rugged range of hills by a very stony track, obtained from the summit a fine view of the well-cultivated plain of Monjenou, bounded by the lofty mountain of Mungasht, now covered with snow. The high hills to the south of this plain were considered the boundary of the Bakhtiyari country. They are inhabited by the Bahmei, a

[2] *I.e.* the king's garden.

branch of the great tribe of Kuhghelu, one of the most savage and lawless in Luristan. Their chief was at his castle of Kala Ala, at some distance from our road, and near the source of a stream bearing the same name (Ab-Ala), which joins the Jerrahi, a river falling into the Persian Gulf. The plain through which we passed had been of late so much exposed to their depredations, that many villages in it had been abandoned by their inhabitants.

Although Mehemet Taki Khan had succeeded in bringing the Bahmei under his authority, and had more than once inflicted punishment upon them for their misdeeds, we should have run some risk had we met one of their 'chapows,' or party of horsemen out on a foray. We had, therefore, to be on our guard during our journey in the wild and deserted country through which we had to pass. A road had anciently been carried through the hills, and we came upon the ruins of an archway that had once crossed it, called 'Getchi-Dervoisa,' or the Limestone Gate. It appeared to have been part of a large building—either a toll-house, or a fort for the defence of the pass. It is probably of the Sassanian epoch, although tradition assigns it to a much earlier age—the Bakhtiyari calling it 'Rustem's toll-house.' My companion pointed out near it an excavation in the rock, which he gravely assured me had been the manger of the renowned horse of that hero of Persian romance, and a tree about fifty yards distant from it to which the animal's hind legs had been tethered.

We descended from these barren hills into the rich and well-cultivated plain of Meï-Daoud, then covered with green crops. It was inhabited by a Bakhtiyari tribe called Mombeni, whose chief, Mulla Fezi, was known as the ' kalunter.' We could see in the distance his castle, on the river Ala, but did not go out of our way to it. We passed many ruins, apparently of the Sassanian epoch, which showed that this part of the province of Khuzistan must anciently have been thickly peopled. The hills surrounding this plain abound in white gypsum, which the Bakhtiyari call 'getchi-oina' (looking-glass limestone).

Another range of low hills separated us from the plain of Ram Hormuz, corrupted by the Lurs into 'Rumes.' From its summit, which we reached at sunset, I obtained a glorious view over the vast alluvial plains which extend to the Shat-el-Arab, or united waters of the Euphrates and Tigris. They were inhabited by nomad Arabs of the tribe of Cha'b.[3] The villages which we could see

[3] The name is written Ka'b, but the Arabs of Khuzistan pronounce the K as Ch in this and other words.

beneath us were surrounded by the graceful palm—a tree that I had not seen since leaving the neighbourhood of Baghdad. The road across these hills —now only an ill-defined and precipitous track— was once defended by a castle, the ruins of which still exist. On our way we passed some naphtha springs, and I heard of others in this part of the country. It was dark before we reached the village of Ram Hormuz, on the river Ala—at this spot a considerable stream. We spent the night, with several other travellers, in the porch of the castle gate.

Ram Hormuz was a celebrated Sassanian city, where Manes, the founder of the Manichæan sect, was put to death by King Behram, and his skin hung up as a warning to his disciples. Its site is marked by numerous mounds which surround the present village. We passed through the midst of them, but I did not perceive any ruins of buildings above ground. The plain is exceedingly fertile, but was ill-cultivated. The chief of the tribe which inhabited it, one Abd'ullah Khan, lived in the small castle of Deh Ure. We did not stop there, but rested for the night at the village of Juma. About two miles distant from it there was a small white-domed Imaum Zadeh,[4] surrounded by palms and

---

[4] The name given to the shrines, or buildings raised over the tombs, of Musulman saints.

orange trees, which contained the tomb of a saint held in great veneration. The Bakhtiyari bring their dead from a distance to be buried near it. We found some men washing a corpse previous to its interment.

We followed the banks of the Ab-Ala through a thick jungle, from which we roused many wild boars, and what appeared to me to be large jackals, but which my companion declared were 'sag-gourgs' (dog-wolves), and, according to him, an altogether different animal. We passed during the day through a country which, in consequence of the depredations of the Bahmei tribe, had been almost reduced to a desert. The population had fled, leaving their villages to fall to ruins. The inhabitants of Joizou, where we spent the night, were of that tribe, and a most ill-looking set of ruffians. In the chief's house I saw a chair and several articles of attire which had evidently once belonged to a European. The chief alleged that they had been the property of a 'Feringhi' (Frank) who had visited the village many years before and had died there. They had more probably belonged to some unfortunate traveller who had fallen a victim to these notorious robbers and cut-throats.

Next day we reached Behbahan, situated in an extensive plain separated from that of Ram Hormuz by hills of limestone and gypsum. The streams which descend from them, such as the Jerrahi, called in this part of its course the Kurdistan, are in consequence brackish and undrinkable. The town is about three and a half miles in circumference, and is surrounded by a mud wall with equidistant circular towers and bastions. Its castle, known as 'Kala Naranj' (castle of the orange), has lofty mud walls, and is protected by a deep ditch. The place once contained a considerable population, but the constant tribal wars in which its inhabitants had been long engaged, together with the plague and bad government, had reduced it to little more than a heap of ruins. At the time of my visit it could scarcely have held four thousand souls. In its bazars little else but the produce of the country and a few European cotton goods were to be found.

The chief, Mirza Koma, was absent at Bender Dilum. I took advantage of a day's rest to go to the bath; my companion, Mehemet Ali Beg, to get helplessly drunk. From Behbahan the country is broken into low hills, and falls gradually in a series of table-lands to the Persian Gulf. These hills are also of limestone and gypsum, and the springs and the pools of rain-water found in them for the most part brackish. Between Behbahan and the sea there is another plain—that of Zeïtun—the principal village in which is Kala

Cham. At a short distance from it is the castle of Gul ve Gul Ab, celebrated in local history, near which two streams, the 'Ab-shur' (brackish water), and 'Ab-shirin' (sweet water), unite and form a river of some size, known as the 'Zokereh,' or 'Hindyan.' Kala Cham was at one time a considerable village, but it had been depopulated by the plague some years before my visit. The plain is renowned for the excellent quality of its rice.

We spent the night at Kala Cham with Mirza Aga, the governor, and an uncle of Mirza Koma. He was a seyyid, and insisted upon entering into a religious discussion with me, which, however, he conducted fairly and good-humouredly, and without any show of fanaticism. Next day we had a ride of six farsaks to Bender Dilum, crossing low but rugged hills, and following and fording the Zokereh. The brushwood on its banks swarmed with the 'duroj,' or black partridge.

Mirza Koma was lodged in a small mud fort, and the town was filled with his horsemen and matchlock-men. He was a man of polished manners and of an amiable disposition, and although a seyyid, and consequently of Arab origin, not a fanatic, as are most descendants of the Prophet in Persia, but liberal in his opinions. His government was described to me as mild and just. He sought to restrain the marauding habits of the tribes under his rule, and to encourage them to settle in villages and to engage in agriculture. His title of Mirza is a corruption of ' Mir-Zadeh '—born of an emir or prince—and is that generally assumed in Khuzistan by seyyids of distinction. I delivered the letter for him given to me by Mehemet Taki Khan. He received me cordially, and ordered a small sailing-vessel to be at once got ready to take me to Karak.

Mirza Koma was then engaged in an expedition against Bushire, with the object of possessing himself of that place, and of reinstating a certain Sheikh Hussein, its former chief, who had been expelled by the inhabitants and who had taken refuge at Behbahan—hoping thus to add this town to the other territories under his rule. With this object he was desirous of obtaining some old guns which were at Karak when the English took possession of the island. He claimed them as having belonged to Sheikh Hussein, and asked me to be the bearer of a letter to the British authorities there, requesting that he might be allowed to bring them away.

At sunset I went down to the shore and found a very rude and crank boat, manned by four half-naked Arabs, ready to receive me. The ' nâ-khudâ,' or captain, said that we should reach Karak next day. I did not, therefore, take any other provisions with me than some bread and a few pomegranates. The wind was light and favourable, and we set our one large sail with the prospect of having a quick passage. But in the night it came on to blow from the southward, and a high sea soon arose. The 'nâ-khudâ' seemed to lose his head, and we were in some danger of foundering, owing to the leaky and rotten condition of the vessel. We beat about the whole of next day, making little progress. With the south wind there came a heavy downfall of rain. There was a kind of hold, in which were stored rice, fruit, and other produce for sale in the bazar established by the English troops in Karak. I obtained some protection from the storm in it, but my quarters were far from comfortable. The 'nâ-khudâ,' finding that I had no provisions with me except the pomegranates and bread, offered to cook me some rice with dried shark's flesh, grated or pounded, and very much like sawdust in taste and appearance. The mess he made me was not savoury, but seemed to be the usual food of the Arab sailors. I was hungry, and did not refuse it. The water which I was given to drink from a tub was absolutely repulsive.

Fortunately the wind fell as the sun went down. A brisk northerly breeze sprang up, and on the following morning we anchored off Karak.

I at once disembarked at the small landing-place which had been constructed for the use of native provision boats, and passing the sepoy sentinels, made my way towards a house over which the British flag was flying. As I had conjectured, it was that of the chief authority in the island, Colonel Hennell, Resident of the East India Company at Bushire, who had left that port when the British mission was withdrawn from Tehran, and was in charge of the camp at Karak.

Dr. Mackenzie, an army surgeon whom I had met at Baghdad, offered me a bed in his temporary hut. My first thought was a bath, as the Arab boat in which I had spent so many hours was swarming with vermin.

A station had been formed at Karak for our Indian troops, who were kept there until the danger of war with Persia had passed. The English officers lived in small houses built after the Indian fashion, with verandahs and thatched roofs; the men in huts constructed of reeds and sun-dried bricks. There was a village on the island, consisting of a few miserable hovels inhabited by poor fishermen. Since the English occupation an extensive native bazar had been opened, and was well supplied with provisions, such as meat, poultry, eggs, vegetables, and fruit, from the opposite coasts of Persia and Arabia, the natives of which, finding that they were paid in ready money, were eager to bring supplies to our market. At the time of my visit the climate was delightful; but in summer the heat, I was told, was almost unbearable. Fever and other diseases then prevailed, and the troops suffered greatly. The island is a barren rock, and only supplied with water from rain collected in artificial reservoirs.

I remained about a fortnight at Karak, making many agreeable acquaintances among the officers, and spending my time very pleasantly in their society and in that of Colonel Hennell. On Christmas Day I dined with Commodore Brucks, who commanded the squadron of the East India Company's navy in the Persian Gulf. His flag was hoisted on the 'Coote,' a corvette on which I had passed a day as a boy, some eight years before, when she was in the Thames.

During my stay in the island I was under the care of Dr. Mackenzie, for the intermittent fever from which I had suffered so constantly and severely during my journey in Persia and my residence in the Bakhtiyari Mountains. I felt restored to almost perfect health when the time came for my departure. Dr. Mackenzie also supplied me with fresh medicines, of which I was much in need, as my little stock was almost exhausted. At the same time he gave me directions for the treatment of my patient, Au Kelb Ali, Mehemet Taki Khan's brother, whose case, however, he judged from my description to be a hopeless one.

I was anxious to introduce vaccination among the Bakhtiyari, as small-pox was prevalent in their country. They had no remedies for it, and were entirely ignorant of either inoculation or vaccination. Through Dr. Mackenzie I obtained some vaccine lymph. I was greatly pleased to be thus able to confer an important benefit upon my mountain friends, and determined to vaccinate Mehemet Taki Khan's children immediately after my return to Kala Tul, if he would permit it, and thus set an example to others.

The information that I obtained from Colonel Hennell led me to infer that the suspension of diplomatic relations with the Shah and the occupation of Karak were not likely to lead to war, but that the English and Persian Governments would probably come, ere long, to an arrangement. I could not, therefore, encourage Mehemet Taki Khan to look to any support from England in his designs for establishing his independence. But I had grounds for hoping that a trade might be opened between the province of Khuzistan and the Bakhtiyari Mountains and India and Europe.[5] It

[5] Colonel Hennell informed me, in a letter of September 1841, that

was only, however, at Baghdad that there were English merchants to whom I could submit my views on the subject, and who might be disposed to venture upon commercial undertakings in a country notorious for the lawlessness of its inhabitants, in which there was no regularly constituted authority, and of which so little was then known.

Colonel Hennell admitted that the rusty and useless cannon which had been found on the island at the time of its occupation by British troops belonged to Sheikh Hussein, the legitimate chief of Bushire, and offered no objection to their removal by Mirza Koma's agent. Having thus concluded all my business, and having enjoyed some much-needed rest, I prepared to return to Kala Tul.

I took passage for Bender Dilum in an Arab boat similar to the one in which I had crossed to Karak. After beating about against a north wind for many hours without making any way, we had to put back, and it was not until January 7 that I finally left the island. But the following morning, when we were in sight of our destination, the wind fell altogether and we lay motionless for twenty-four hours in a dead calm. I was again obliged to have recourse to the shark pillau of

' although the Government would have nothing to do with Mehemet Taki Khan's political views, he did not think it was altogether indisposed to meet his commercial projects.'

my 'nâ-khudâ.' A breeze which sprang up at noon next day enabled us to reach Bénder Dilum about sunset. I was hospitably received by Haji Aga, the brother of Haji Hassan, the 'lala' or tutor of Mehemet Taki Khan's children, and I passed the night in his humble hut.

I was informed by my host that Mehemet Ali Beg, tired of waiting for me, and probably imagining that having once rejoined my countrymen I should not be disposed to leave them again, had left Bender Dilum for Kala Tul, taking my horse with him. A report had reached him, moreover, that the Matamet had already set out on his expedition against Mehemet Taki Khan with a large army. As he was one of the principal and most trusted retainers of the Bakhtiyari chief, and was expected to be with him in times of difficulty and in war, he considered it his duty to lose no time in returning to his master. I had been detained so long at Karak and by adverse winds that, although his departure had placed me to great inconvenience, I could scarcely be surprised at it.

Mirza Koma having learnt that the inhabitants of Bushire were not favourable to the return of Sheikh Hussein, and that he could not rely upon their co-operation, had renounced his intention of attacking the town, and was on his way back to

Behbahan. I decided upon following him, as without his assistance and protection I should have great difficulty in reaching Kala Tul, especially as, in consequence of the rumours of war, the country through which I should have to pass was already in a very disturbed state. I hoped to be able to join him in a few hours, as he was moving slowly with his tents, irregular cavalry, and numerous camp-followers. But I was unable to hire a horse, and it was only with some difficulty that I procured a donkey to carry me and my saddle-bags. My progress was consequently slow, and I could only cross the sandy belt bordering the sea to the small Arab village of Liletain, where I passed the night.

Next day I was able to procure a horse in the small village of Hussor, after a long and fatiguing ride upon my jaded ass. At sunset I reached Mirza Koma's tents. He received me very cordially, invited me to accompany him as his guest to Behbahan, and promised to assist me in returning to Kala Tul. The village of Ghenowa, where he was encamped, was filled with his horsemen and 'tufungjis' (matchlock-men). As the weather was delightfully warm, I passed the night on my small carpet in the open air.

It was January 25 before we reached Behbahan, as we were detained several days by heavy rains.

The baggage animals had great difficulty in crossing the swollen torrents. A part of the plain had become a morass. To the great amusement of Mirza Koma, I sank on one occasion in a quagmire, from which I was dragged with my horse, not without some trouble, by his attendants. The country was already carpeted with flowers, and the jonquil and the narcissus—the Persians call it 'nerkis'—filled the air with the most grateful perfume. The plains and valleys of Behbahan deserve their reputation of being one of the 'bihishts,' or paradises, of Persia. The horsemen of Mirza Koma were constantly dismounting and gathering handfuls of narcissus, with which they adorned themselves and their horses. The chief himself would, every now and then, direct his carpet to be spread on a flowery bank near some stream, and invite me to smoke a kaleôn and to drink sherbet with him. Although we were in the month of January, the air was warm and balmy.

The Mirza, being a descendant of the Prophet and of a distinguished seyyid family, was preceded on his march by a large flag of green silk embroidered in gold with texts from the Koran. The standard-bearer was accompanied by musicians on horseback, beating drums and playing on a kind of oboe. The Mirza himself was escorted by some five hundred horsemen. He and many of his retainers rode handsome high-bred Arab mares. Some of the chiefs had with them their hawks and hounds—hunting and war going together—and scoured the plains in pursuit of game, with which they abounded. The principal sport consisted in capturing the 'houbara,' or middle sized bustard. This bird is taken by a large falcon called 'chalk,' trained for the purpose. When it is frightened by the approach of the horsemen, it endeavours to escape by running or by concealing itself in the long grass. The falcon—released from its hood and raised high on the wrist of the sportsman—soon perceives its quarry, and skimming rapidly along the ground, rises on approaching it, and, without hovering above it, strikes it at once. The bustard rarely attempts to evade its enemy by flight, but usually makes a gallant resistance, in which it sometimes proves victorious. The horsemen, to prevent their hawks from being injured, ride up at once, separate the combatants, kill the bustard, and reward the falcon with its victim's brains.

We captured a great number of 'duroj,'[6] red partridges, ducks, and other game birds, and the greyhounds coursed gazelles and hares. Game was very plentiful in the plains through which we passed, owing to their want of population.

[6] Or francolin, the black partridge of India.

On approaching the villages the inhabitants came out to meet the Mirza, the women making the loud vibrating noise, called by the Arabs the 'tahlel,' by striking their mouths rapidly with the palms of their hands whilst uttering a shrill cry.

The principal inhabitants of Behbahan had left the town to meet their governor the day before his entry. He had encamped near a spring called the 'Chahi-Wali' (the Wali's well),[7] at a short distance from his capital, and the omens had designated the following day for an auspicious return to it. A crowd of men, on horseback and on foot, were assembled outside the walls to receive him. They crowded to kiss his hand, as he was looked upon, on account of his descent from the Prophet, as a sacred personage. The horsemen galloped over the plain, engaging in mimic fight. Every one who carried a gun fired it off, and we passed through the gate amidst the shouts of the population and salutes of artillery. The houses were adorned with flags and coloured hangings, and their flat roofs covered with women making the 'tahlel.'

The Mirza stopped at the entrance of the principal mosque and repeated a short prayer, whilst an

[7] Wali was the title formerly given to the governor of the district of Behbahan.

almost naked dervish called down blessings upon his head in a stentorian voice. We then rode to the castle. The chief entered his enderun immediately after his arrival, and I was left to myself in a small room which had been assigned to me. To my great relief I found that Mehemet Taki Khan, convinced that I would keep my promise of returning to Kala Tul, had sent a man with a horse to meet me as soon as Mehemet Ali Beg had arrived there with mine.

I found, as I travelled towards Kala Tul, that the state of the country had changed considerably since I had passed through it only a short time before. Mehemet Taki Khan, anticipating the invasion of his mountains by the Persians, had summoned all the horsemen and matchlock-men of the tribes to join him, and the Bahmei, taking advantage of their absence, were plundering the villages and driving off the sheep and cattle. The road was consequently very unsafe, and I left Behbahan accompanied by ten armed men, whom my guide considered necessary for my protection. This borderland between the lawless mountain tribes of the Kuhghelu, the Cha'b Arabs, and the Bakhtiyari, is at all times subject to their depredations, and a very fertile district thus remains almost uninhabited and waste.

As the villagers feared to leave their homes,

my guide was unable to procure an escort beyond the village of Sultanabad. Dreading lest he should fall into the hands of the enemies of his tribe, and thinking that he would have a better chance of getting safely through the dangerous tract without my company, he made off in the middle of the night, and left me to shift for myself. The inhabitants of Sultanabad informed me that on the previous day a body of Arab horsemen, commanded by the son of a certain sheikh Moslet, a notorious robber, had made a raid in the plain of Ram Hormuz, and had driven off cattle and sheep. I was earnestly warned against the danger of falling into the hands of these marauders, and of being robbed, if not murdered, were I to proceed alone. But I could not remain in the village for an indefinite time, and hoping for the best I went on my way. After I had ridden for some time without meeting any one, I perceived in the distance a body of horsemen. I made up my mind that they were the Arabs who had been pillaging the district. As they must have seen me as soon as I saw them, and as it would have been utterly useless for me to attempt on my tired horse to escape from them, I decided to advance to meet them, and to make myself known. Fortunately, they proved to be Arabs under the command of a Cha'b sheikh named Ahmed, who, in the absence of Abd'ullah

Khan, the chief of Ram Hormuz, had come to assist his people in defending themselves and their property.

The unsettled state in which this part of Khuzistan had long been, arose from the wars between Mehemet Taki Khan and the chiefs of Behbahan and the Kuhghelu and Mamesenni tribes. The plain of Ram Hormuz at one time belonged to the family of Mirza Koma. The Ferman-Fermai, one of the Persian princes, and governor of Shiraz, had deprived Mirza Mansur, Mirza Koma's eldest brother, of the governorship of the town and district of Behbahan, and had named one of his own sons, known as the Wali, in his stead. The two brothers took refuge with Mehemet Taki Khan, and appealed to him to assist them in recovering their territories. He consented to do so, and marched with a strong force against the Wali, who had fortified himself in the castle of Behbahan. It was invested, and forced, after a short resistance, to surrender for want of provisions. The Wali was murdered as he was leaving the town.

Mehemet Taki Khan reinstated Mirza Mansur, and received as the reward for the services he had rendered him the plain of Ram Hormuz. But it belonged, at that time, to the Arab sheikh Moslet, who had been tributary to the chiefs of Behbahan. He refused to cede his lands to the Bakhtiyari chief, or to pay tribute to him. A war was the consequence, and he was defeated, made prisoner, and put to death. His tribe then moved from Ram Hormuz, and pitched their tents in the plains on the right bank of the river Karun. But they had become the mortal enemies of Mehemet Taki Khan, in consequence of the blood-feud between them, and were constantly making descents upon those who had been settled on the lands which they had been compelled to abandon. Mirza Koma, who had succeeded to the governorship of Behbahan, was suspected to be no stranger to these proceedings, as he was known to be desirous of recovering the fertile district of Ram Hormuz. A year before he had attacked the castle of Kala Sheikh in the plain, and had been defeated in his attempt to possess himself of the place by Au Aslan, the nephew of Mehemet Taki Khan. This had led to unfriendly relations between him and the Bakhtiyari chief.

In the raid which had taken place the day before I passed through the plain, several of the villagers had been killed and wounded. Among the cattle captured were some cows belonging to a seyyid. He followed the Arabs, and appealed to their sheikh, who, out of respect for his sacred character as a descendant of the Prophet, gave them back to him. The old man, who was suffering from inflammation in his eyes, applied to me for medicine. He seemed to derive some benefit from the lotion I gave him, for the following morning he came to express his gratitude, and insisted upon my mounting his horse and accompanying him to his village. After he had served an excellent pillau for my breakfast, he sent one of his sons with me to visit the orange trees in the gardens of Anushirwan. In one of these gardens was an artificial mound which, according to a tradition, covers the remains of a palace of that renowned monarch of the Sassanian dynasty, and near the place I was shown his tomb and that of his son.

In the village of Ram Hormuz I found the man whom Mehemet Taki Khan had sent to conduct me to Kala Tul. I reproached him for his cowardice in leaving me. He excused himself by saying that if he had fallen into the hands of the Arabs, who were expected to make a 'chapou' on that very day in the plain, he would have had his throat cut, as they had a blood-feud with his tribe, whilst if they had taken me they would have done me no harm, but would only have left me naked. We set off together for Kala Tul.

On arriving at the castle I found that Mehemet Taki Khan had already left it, with the chiefs who had joined his standard, his retainers, and the horsemen and matchlock men collected from the tribes. Khatun-jan Khanum and the other inmates of the enderun were delighted to see me. I learnt from her all that had taken place during my absence. The Matamet, finding that he was unable to obtain the money he had demanded from Mehemet Taki Khan as the tribute of the Bakhtiyari tribes, and accusing him of being in rebellion against the Shah, had determined to undertake an expedition against him. Large arrears of taxes were also due from the cities of Shuster and Dizful, and from the Arab population of Khuzistan, and he intended to enforce their payment at the same time. He had already commenced his march, and had entered the mountains by the Zenda-rud and Zerda-kuh. The Shah had commanded Ali Naghi Khan, Mehemet Taki Khan's brother, to accompany the governor of Isfahan as a hostage and as his guide to Kala Tul. Mehemet Taki Khan, uncertain as to the course he should pursue, whether to submit or resist, had gone to Mal-Emir, where he had encamped with his followers. The Matamet would have to descend from the high mountains into that plain, and the Bakhtiyari chief would be able to determine how to act. His wives and his family, and those of his relations and of his adherents, were in great alarm at the prospect of a war and the possibility of an occupation of their country by the Persian troops, who, they knew, would commit every manner of excess and outrage upon them. They were already making preparations to leave Kala Tul, and to seek for safety with their children and property in the almost inaccessible mountains in which the tribe had their ' sardesirs,' or summer pastures.

Khatun-jan Khanum, who had been left in charge of the castle, feared lest the Bahmei tribe and the Arabs under the son of Sheikh Moslet, taking advantage of Mehemet Taki Khan's absence from Kala Tul, would plunder the inhabitants of the country in its neighbourhood, as they had already done those of the plain of Ram Hormuz. It was rumoured that a 'chapou' party had been seen at no great distance from the village, and much alarm was felt lest it should be attacked, as it was without sufficient means of defence. The Khanum, therefore, decided to send out as many horsemen and matchlock-men as could be collected together, under the command of Au Azeez, one of her relations, to reconnoitre and to hold the enemy in check. I accompanied the young chieftain. We concealed ourselves during the day in the low hills beyond Manjanik, and resumed our march after nightfall.

It was scarcely dawn when we saw in the distance a company of horsemen. We could not at first make out whether they were marauders, or peaceful traders on their way to Shuster. My companions, keeping out of sight in a ravine, made preparations to fall upon them. Hidden behind a rock, I watched the party as they drew near, and thought that I perceived among them a European wearing a cap with a gold-lace band. I begged my Bakhtiyari friends to remain concealed until I could ascertain who this European might be. Approaching him alone, I called to him in French. He was not a little surprised at being addressed in that language by a Bakhtiyari, for whom, on account of my dress, he at first mistook me. I found him to be the Baron de Bode, whose acquaintance I had made in the Shah's camp at Hamadan. He was accompanied by an escort of irregular horse, which had been furnished to him by the Persian authorities, and had a train of servants and baggage mules. He informed me that he was on his way to join the Matamet, of whose movements I was able to give him some information.

I returned to my companions and warned them of the danger of attacking and robbing a secretary to the Russian Embassy. If he happened to be killed in the affray the Russian Government would, I said, inevitably insist upon redress, and the consequences might prove very serious to Mehemet Taki

Khan and his tribe. They acted on my advice, and allowed the Baron to pass unmolested—still, however, remaining concealed in the ravine. It was only some years after, when I met him in a London drawing-room, that I informed him of the danger which he had run; for had I not restrained Au Azeez and his followers, they would probably have fired a volley into his party, which would have had fatal results.

As we saw no enemy, and had consequently reason to believe that there was no ground for the alarm of the inhabitants of Kala Tul, we returned in the course of the day to the castle.

I soon joined Mehemet Taki Khan at his camp in the plain of Mal-Emir, taking with me his two eldest boys, whom their mother committed to my care. He informed me that he had determined not to oppose the passage of the mountains by the Matamet, but to receive him as a guest, and by protesting his loyalty and subjection to the Shah to endeavour to avoid a conflict. He hoped to conciliate the Governor of Isfahan by presents, and by the payment of so much of the tribute claimed from him as he might be able to collect from the tribes which recognised his authority. Consequently his brother, Ali Naghi Khan, was not only acting as a guide to the Persian army through the mountains, but the Bakhtiyari on the way had received orders to help in the transport of the guns, which the Persian artillerymen, without their assistance, could not have dragged over the steep and rocky passes of the great range which separates the centre of Persia from the province of Khuzistan.

Mehemet Taki Khan's camp occupied a large area. It was composed of the usual black tents and of huts constructed of reeds and boughs of trees. He had collected a force of about eight thousand men, including horsemen and men on foot armed with matchlocks. Most of the tribes acknowledging his authority, including the Arabs from the banks of the Jerrahi and from the plains around Shuster, had furnished their contingents. A more motley and a wilder and more savage set of men it would have been difficult to bring together. They were very warlike in their demonstrations, constantly firing off their loaded guns, to the great danger of those who might be near, dancing their war-dance and shouting their war-songs. They only awaited a word from Mehemet Taki Khan to fall upon the Matamet and his regular troops. Encumbered as these were with artillery, baggage, and the usual following of a Persian army, in the difficult mountain passes and narrow defiles, they might easily have been cut to pieces.

I frequently witnessed whilst in Mehemet Taki Khan's camp the effect which poetry had upon men who knew no pity and who were ready to take human life upon the smallest provocation or for the lowest greed. It might be supposed that such men were insensible to all feelings and emotions except those excited by hatred of their enemies, cupidity, or revenge. Yet they would stand until late in the night in a circle round Mehemet Taki Khan, as he sat on his carpet before a blazing fire which cast a lurid light upon their ferocious countenances—rather those of demons than of human beings—to listen with the utmost eagerness to Shefi'a Khan, who, seated by the side of the chief, would recite, with a loud voice and in a kind of chant, episodes from the 'Shah-Nameh,' describing the deeds of Rustem, the mythical Persian hero, or the loves of Khosrau and Shirin. Or sometimes one of those poets or minstrels who wandered from encampment to encampment among the tribes would sing, with quavering voice, the odes of Hafiz or Saadi, or improvise verses in honour of the great chieftain, relating how he had overcome his enemies in battle and in single combat, and had risen to be the head of the Bakhtiyari by his valour, his wisdom, his justice, and his charity to the poor. The excitement of these ruthless warriors knew no bounds. When the wonderful exploits of Rustem were described—how with one blow of his sword he cut horse and rider into two, or alone vanquished legions of enemies — their savage countenances became even more savage. They would shout and yell, draw their swords, and challenge imaginary foes. When the death of some favourite hero was the poet's theme, they would weep, beat their breasts, and utter a doleful wail, heaping curses upon the head of him who had caused it. But when they listened to the moving tale of the loves of Khosrau and his mistress, they would heave the deepest sighs—the tears running down their cheeks —and follow the verses with a running accompaniment of 'Waï! waï!'

Such was probably the effect of the Homeric ballads when recited or sung of old in the camps of the Greeks, or when they marched to combat. Such a scene as I have described must be witnessed to fully understand the effect of poetry upon a warlike and emotional race.

Mehemet Taki Khan himself was as susceptible to it as his wild followers. I have seen him when we were sitting together of an evening in the enderun at Kala Tul, cry and sob like a child as he recited or listened to some favourite verses. When I expressed to him my surprise that he, who had seen so much of war and bloodshed, and had himself slain so many enemies, should be thus moved to tears by poetry, he replied, 'Ya, Sahib! I cannot help it. They burn my heart!'

The shrill notes of a kind of oboe, not unlike those of a Scotch bagpipe, and the monotonous beat of the drum, were heard night and day in the tents of the Bakhtiyari. They appeared to afford as much delight and to cause almost as much excitement to these wild mountaineers as their beloved poetry.

END OF THE FIRST VOLUME.

PRINTED BY
SPOTTISWOODE AND CO., NEW-STREET SQUARE
LONDON

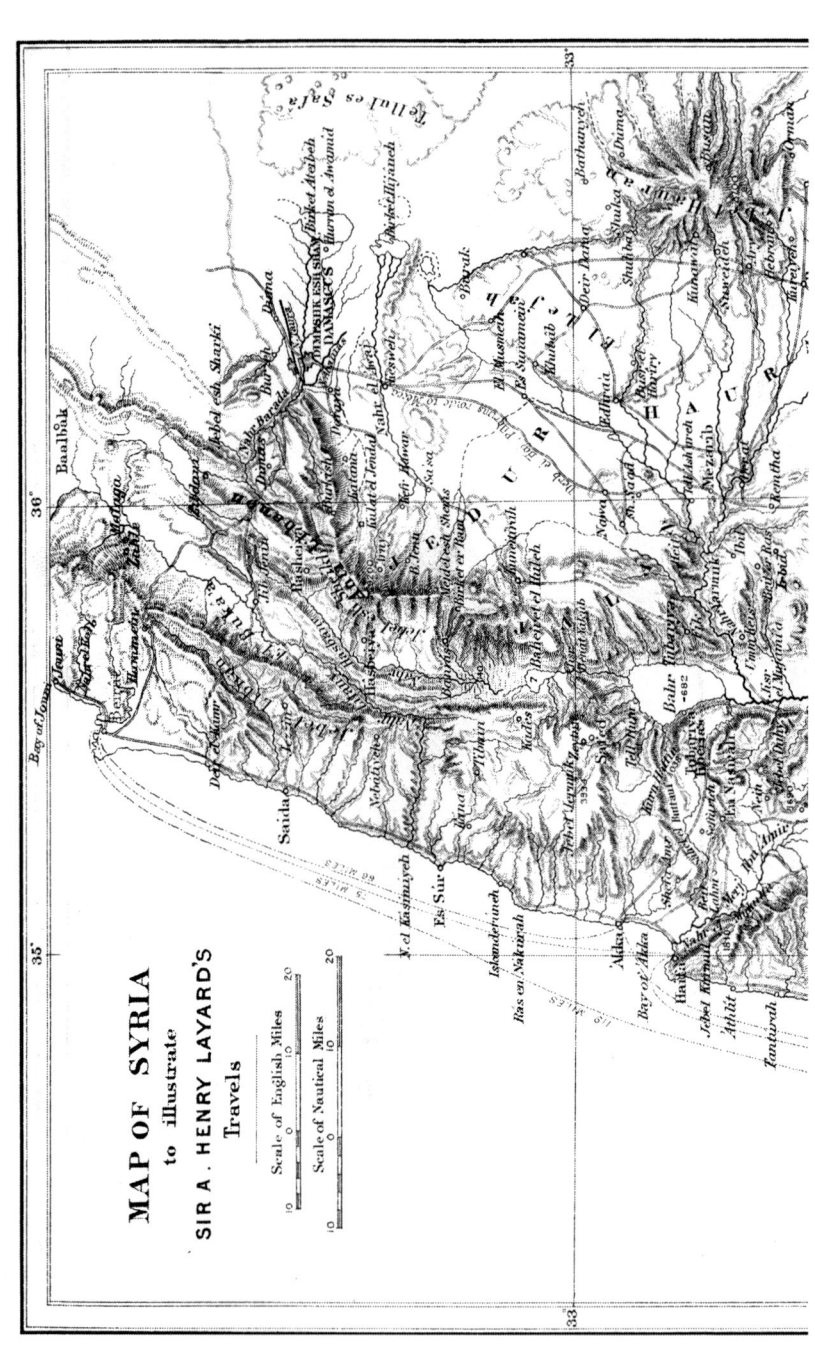

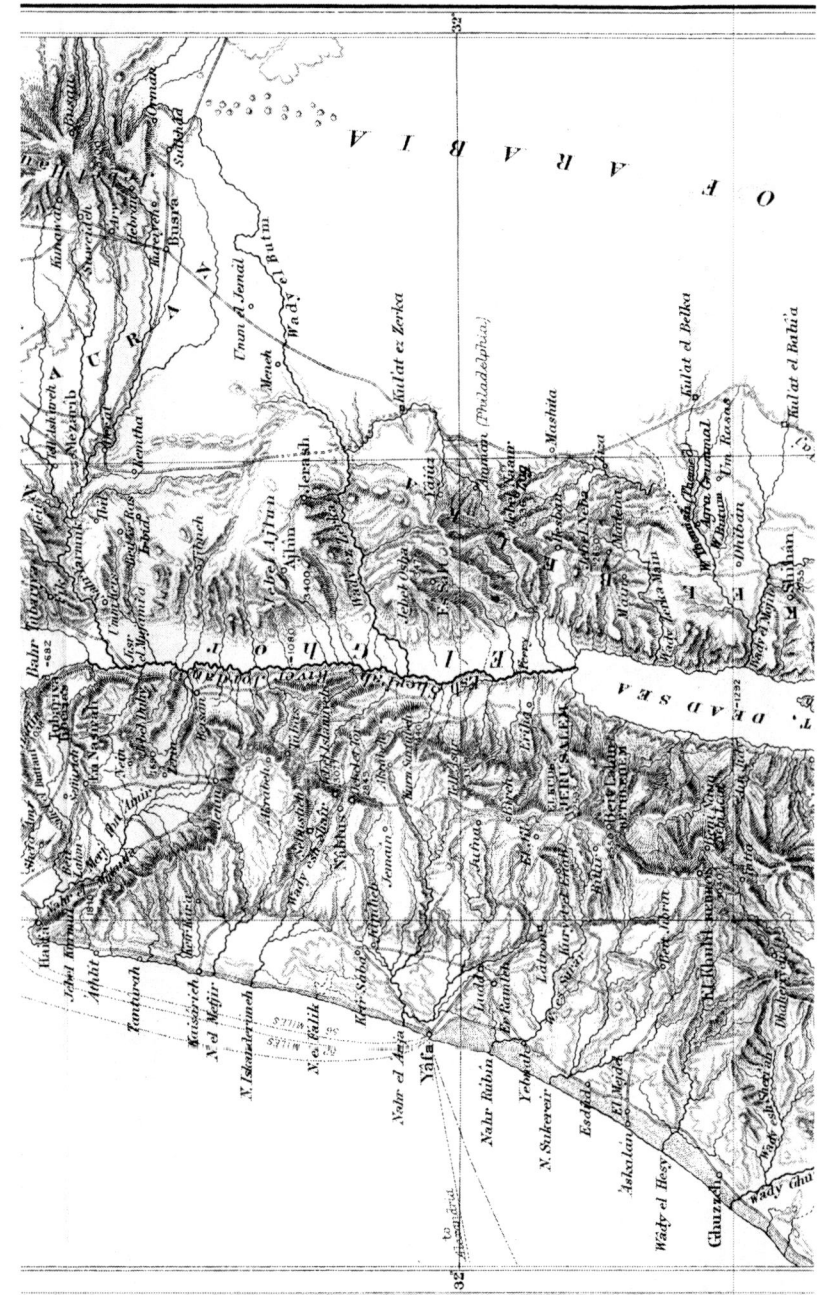

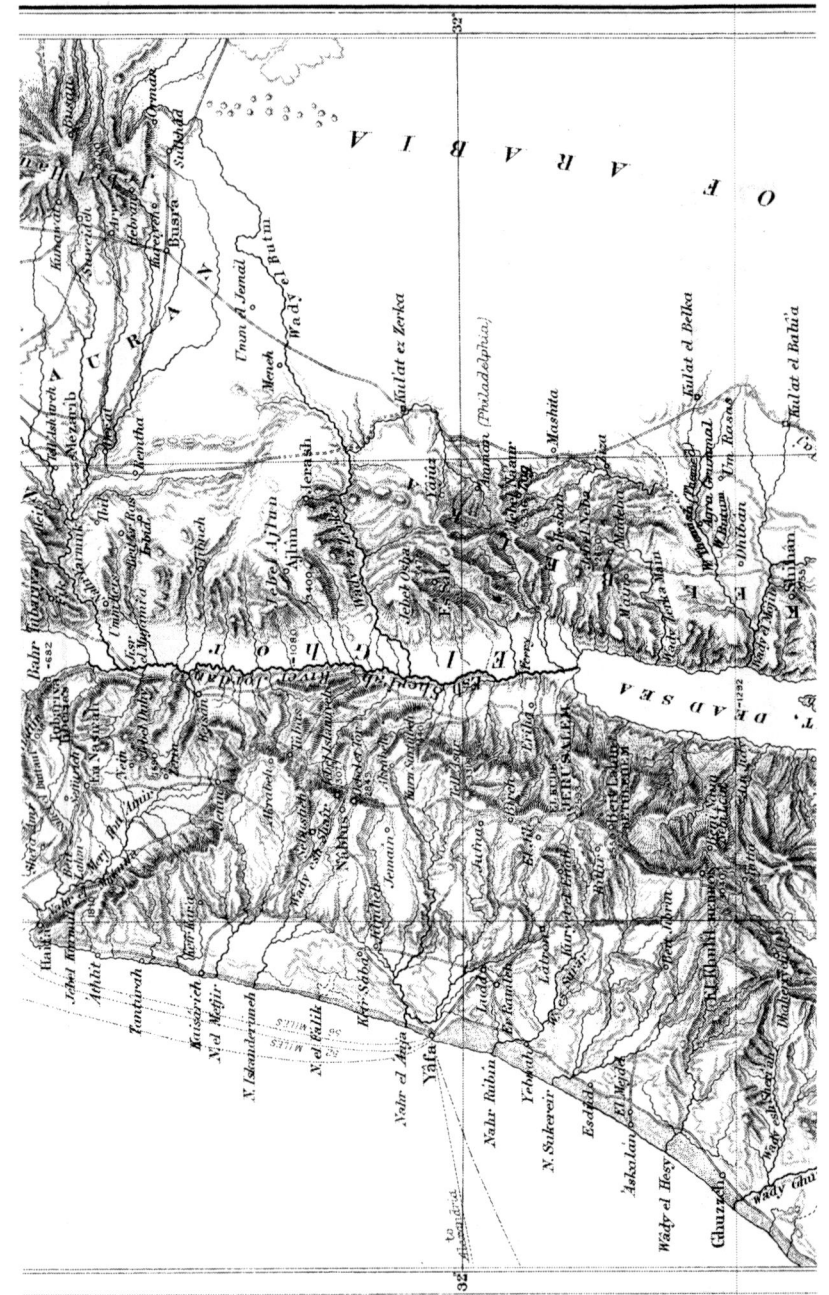

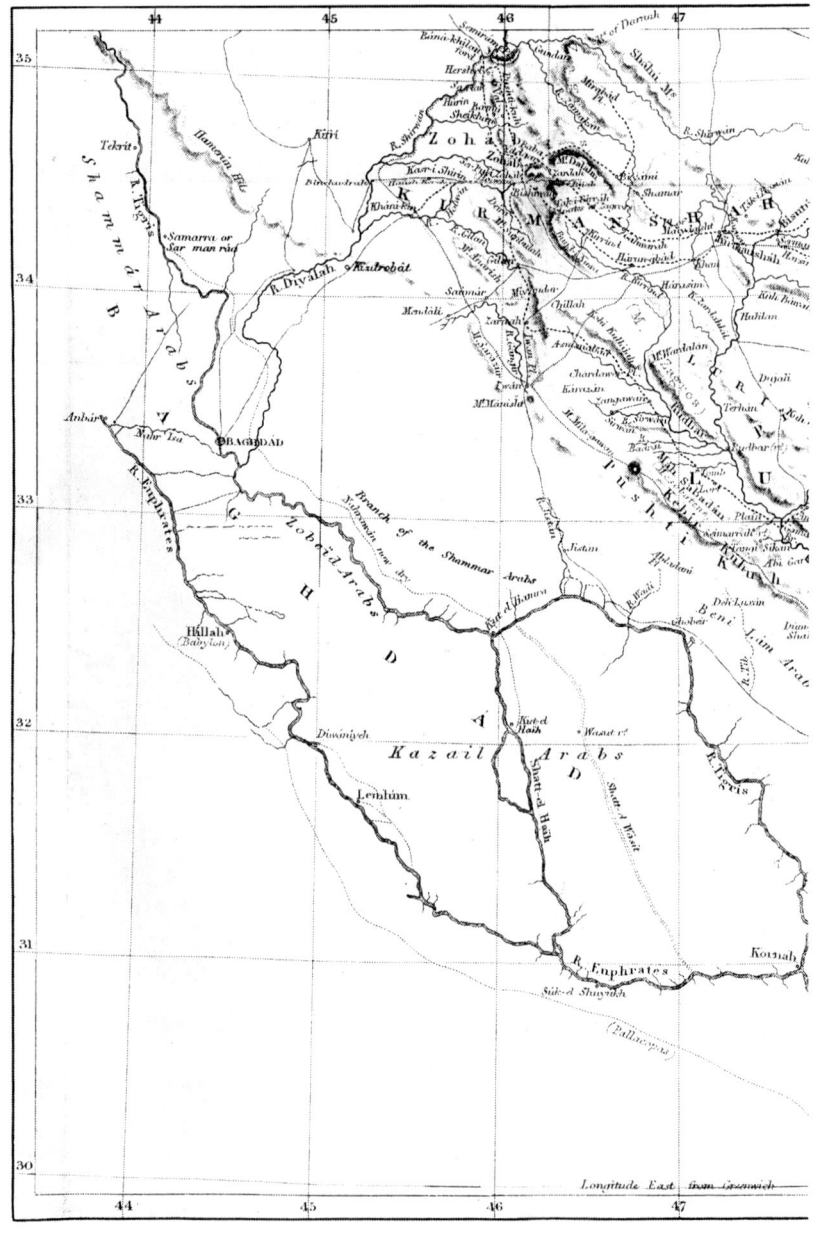

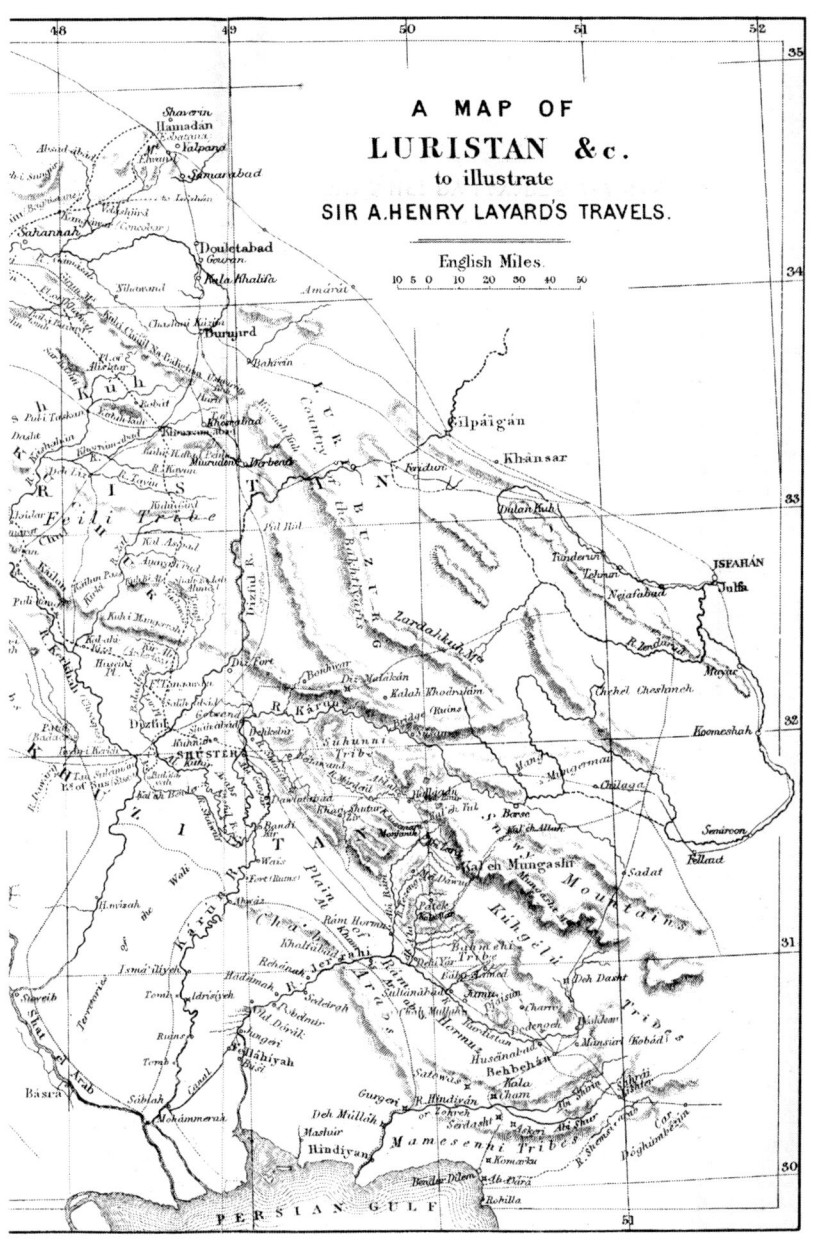

Lightning Source UK Ltd.
Milton Keynes UK
UKOW04f0638070115

244077UK00001B/49/P